Eighteenth-Century Geography and Representations of Space

Jean-Paul Forster

Eighteenth-Century Geography and Representations of Space

in English Fiction and Poetry

PETER LANG
Bern · Berlin · Bruxelles · Frankfurt am Main · New York · Oxford · Wien

Bibliographic information published by die Deutsche Nationalbibliothek
Die Deutsche Nationalbibliothek lists this publication in the Deutsche Nationalbibliografie; detailed bibliographic data is available on the Internet at ‹http://dnb.d-nb.de›.

Library of Congress Cataloging-in-Publication Data

Forster, Jean-Paul.
Eighteenth-century geography and representations of space in English fiction and poetry / Jean-Paul Forster.
pages cm
Includes index.
ISBN 978-3-0343-1257-8
1. English literature–18th century–History and criticism. 2. Space and time in literature. 3. Geography in literature. I. Title.
PR448.S69F67 2013
820.9'32–dc23

2013020705

Cover illustration: 2 details from John Ogilby's ribbon maps in his atlas *Britannia*. By kind permission of Theatrum Orbis Terrarum, publishers.

ISBN 978-3-0343-1257-8 pb.   ISBN 978-3-0352-0211-3 eBook

Hochfeldstrasse 32, CH-3012 Bern, Switzerland
info@peterlang.com, www.peterlang.com

Printed in Switzerland

For Edith

Acknowledgments

I would like to express my deep gratitude to my first reader, Anthony Pereira, for his patient and pertinent comments on the manuscript. This book owes its origin to a happy coincidence: the convergence of my interest in geographical representations of space and the subjects chosen by the *Centre de recherche et d'études anglaises* of the Sorbonne Nouvelle for their international conferences. For the years 2004–2007, these subjects were the sea and motion. What follows will amply show my indebtedness to these interdisciplinary colloquiums held in memory of the late lamented Paul-Gabriel Boucé. I seize this opportunity to thank the joint presidents of the *CREA XVIII*, Suzy Halimi and Serge Soupel and the participants for enriching this reflexion on geographical space. I must also mention the assistance I received from various Swiss libraries and their staff, in the first place the staff of the Bibliothèque cantonale et universitaire in Lausanne, who have been extremely generous with their time, the Zentral Bibliothek in Bern, and Mrs Nathalie Hutter of the University library of St Gall.

Contents

List of Illustrations

5.1 William Pars, "Bridge near Mount Grimsel". Watercolour. © The Trustees of the British Museum. London.

5.2 John Robert Cozens, "Chamonix and Martigny–The Aiguille Verte", known today as l'Aiguille de Dru. © Victoria and Albert Museum. London.

5.3 William Pars, "Mer de Glace above Chamonix". Watercolour. © The Trustees of the British Museum. London.

5.4 J.M.W. Turner, "The Devil's Bridge. The St Gotthard Pass". © Tate. London.

5.5 William Henry Bartlett, "The Summit of the Jungfrau. Scene from *Manfred*". By courtesy of Viaticalp, database of the Bibliothèque cantonale et universitaire in Lausanne.

5.6 J.M.W. Turner, "The Bay of Baiae, Apollo and the Sybil". © Tate. London.

5.7 J.M.W. Turner, "Snow Storm–Steamboat off a Harbour". National Gallery. © Tate. London.

Introduction

Literature is the product of perception, or, more simply is perception.

Douglas C. D. Pocock[1]

La sensation n'est jamais pure: l'individu vit dans une société, utilise un vocabulaire de formes et de couleurs qui prédéterminent ce qu'il ressent; il perçoit le monde à travers les grilles de lectures qu'il a reçues.

Paul Claval[2]

The place where we live and the space where our activities unfold are determining factors in our lives. Yet for hundreds of years, except for travellers and historians, writers had not been much concerned with the physical aspects of the milieus they described. Their geography had been hazy. Homer's Troy, Virgil's Carthage, King Arthur's Camelot, Shakespeare's or Ben Jonson's Venice, and even Don Quixote's Spain were more mythical or legendary places than real cities, castles and countries. When topographical features were mentioned, like the Charybdis and Scylla reefs in *The Odyssey* or the narrow pass in the Pyrenees where Charlemagne's nephew meets his death in *La Chanson de Roland*, it took the ingenuity of scholars to identify them with precise spots centuries later. All this changed at the turn of the eighteenth century, the century when the search for locating legendary, mythical and biblical places began in earnest: a sign of the times. In the British Isles, fiction and a little later poetry began to make more systematic references to the topography of the earth. At first, a few writers explicitly, others indirectly, claimed in this to be following the example of geographers. Their geography was that of the globe. In works that told their characters' or narrators' travels, the attention paid to topography

1 Douglas C. D. Pocock, *Humanistic Geography and Literature* (London: Croom Helm, 1981), 15.

2 "Sensations are never pure: the individual lives in a society, uses a vocabulary of forms and colours which predetermine what he feels: he sees the world through the grids of his readings." Paul Claval, *La Géographie culturelle* (Paris: Nathan, 1995), 61.

passed mostly unremarked, the fictions being read as adventure stories, fake or true biographies or satires. During the rest of the century, the reference to actual geography, including local topography, went on asserting its presence in literary representations of space. The development had a definite connection with the popularity that geography enjoyed at the time. Maps, atlases and travel reports contributed to giving people a clearer idea of what the surface of the earth looked like. This book looks into what might be called literature's discovery of this modern geography. It examines illustrative representations of space and their contribution to our vision of the world.

What is puzzling is that the public's interest in the overall and systematic description of the surface of the earth proposed by geography did not start earlier: that it occurred when it did. After all, in the British Isles, Hakluyt's and Purchas' collections of accounts of explorations had long aroused English readers' curiosity about remote seas and continents. In Europe, prestigious atlases and maps were furthermore the work of geographers of the previous century: there was no cartographer of Johan Blaeu's stature to make atlases more particularly popular in the eighteenth century. On the contrary, the reproduction of maps of the earth, continents or countries had become a repetitive and unimaginative business. Finally, until James Cook's and Louis Antoine de Bougainville's reports on paradisiacal islands in the South Pacific late in the century, no remarkable geographical discoveries took place that could have suddenly revived a blunted curiosity. To invoke the publication of a spate of records of recent circumnavigations of the globe between the 1690s and 1710s is not fully convincing either. They often made dreary reading compared with earlier accounts and their impact has probably been exaggerated. Defoe himself seems to have thought so when he blamed them for lacking the spice of adventure.[3] As for the story of Alexander Selkirk that inspired his tale of a shipwrecked hero on an uninhabited island, it can only have played a minor role as a catalyst.

The popularity of the geography of the seas and continents can only have been due to a combination of more diffuse factors: the increasing mobility of the population and the expansion of European settlements overseas, the development of trade with them. More people from all

3 *A New Voyage round the World* (1725), ed. John McVeagh, *The Novels of Daniel Defoe* (London: Pickering and Chatto, 2009), 30.

classes travelled for all sorts of reasons, social and economic mainly. Voyages across the Atlantic and the Indian Ocean had become routine. Ships carried emigrants to other continents and brought back prized goods to Europe. Geography was simply more in the news than it had been and interest in the world at large was not limited to the cosmopolitan crowd of businessmen portrayed in Joseph Addison's *Spectator* essay, "The Royal Exchange". Following Edmund Halley's 1676 expedition to St Helena to study the stars of the southern hemisphere, trade winds and monsoons and his 1700 expedition to study the conditions of terrestrial magnetism, the number of ventures of a scientific kind increased. Such expeditions made navigation more efficient. The result was that sea voyages of whatever kind became the talk of London. They were discussed in coffee houses and taverns. Geographical information, including maps, found their way into cheaper publications. Maps, atlases, globes and travel literature were available in bookshops, at least in London. They could also be consulted in or borrowed from circulating or subscription libraries. There is even some evidence that this geographical material was more popular with their patrons than poetry and fiction.[4]

If interest in geography first centred on overseas travels and discoveries, less distant but equally little known areas nearer home attracted curious visitors. Other visitors in turn used their reports as guidebooks. Although written by amateur geographers, guidebooks were accepted as geographical works. When Daniel Defoe and Samuel Johnson wrote, the former his *Tour through the Whole Island of Great Britain* (1724–1727) and the latter his *Journey to the Western Islands of Scotland* (1775), they deliberately acted as geographers. To such publications must be added the first actual guidebooks: guidebooks for the visitors of London, spas and, towards the end of the century, Wales, the Lake District and Scotland. Guidebooks were in fact an invention of the period.[5]

One effect of the general curiosity for things geographical was that allusions to them, and particularly to maps, sometimes crop up in unexpected contexts. Who else but a novelist of the period could have thought of amusing his readers with an idle promise of illustrating one

4 John Brewer, *The Pleasures of the Imagination: English Culture in the Eighteenth Century* (New York: Farrar, Strauss, Giroux, 1997), 172, 174, 177 and 181.

5 This is still a little investigated field in history and literature.

of his characters' local reputation by means of a map, as Laurence Stern does in the case of Tristram Shandy's midwife? And is not the following vindication of geography out of place in the same novel, considering that it stages characters leading a sedentary way of life?

> … as for geography, 'tis of absolute use to [a soldier]; he must be acquainted intimately with every country and its boundaries where his profession carries him; he should know every town and city, and village and hamlet, with the canals, the roads, and the hollow ways which lead up to them; there is not a river or a rivulet he passes, Trim, but he should be able at first sight to tell thee what is its name—in what mountains it takes its rise—what is its course—how far it is navigable—where fordable—where not; he should know the fertility of every valley, as well as the hind who ploughs it; and be able to describe, or, if it is required, to give thee an exact map of all the plains and defiles, the forts, the acclivities, the woods and morasses, thro' and by which his army is to march; he should know their produce, their plants, their minerals, their waters, their animals, their seasons, their climates, their heats and cold, their inhabitants, their customs, their language, their policy, and even their religion.[6]

This vibrant apologia, in which Uncle Toby underlines, for the benefit of his listener, Corporal Trim, the importance of topographical knowledge in the planning and carrying out of military operations, in fact refers to another aspect of eighteenth-century geography besides its general mapping of the globe: its less often mentioned depiction of regional geography, which gave people practical knowledge of an area. This interest of the age in things geographical spread to other fields of studies. Geography claimed the attention of historians in a way it had never done—one thinks here of Defoe's historical writings, of Montesquieu's *Considérations sur les causes de la grandeur des Romains et de leur décadence*, of his *Esprit des lois* and of Edward Gibbon's *Decline and Fall of the Roman Empire*. It even stole into the theological debate, first over the original location of paradise in the Near East and, in the second half of the century, over some of its possible survivals on some blessed remote Pacific islands.[7]

6 Laurence Stern, *The Life and Opinions of Tristram Shandy* (Oxford: Oxford University Press, 1983), 31, 454–55.

7 Charles W. J. Withers, "Geography, Enlightenment, and the Paradise Question", *Geography and Enlightenment*, ed. David N. Livingstone and Charles W. J. Withers (Chicago: Chicago University Press, 1999), 67–92.

However incomplete and imprecise it remained, the corpus of eighteenth-century geographical knowledge offered an image of the earth with which, throughout the period, the public became sufficiently well acquainted so that, eventually, "even if a geographical book lacked maps, those maps were nonetheless available and provided the geographical reader's intellectual framework."[8] Geography's description of the surface of the earth had become a central feature of what has been called by an apt phrase the "horizon of expectation" of the public, namely the knowledge largely available to the people of a period. [9] As it existed predominantly in the form of discourses, one could also speak of a horizon of discourses. Geography and maps were so popular that "mapping" even came to be used as a common metaphor for organizing fields of scientific observations and a model of encyclopaedic empirical knowledge. For governments, trade and the public at large, its descriptions of places were so many invitations to learn more and travel farther, and maps facilitated the planning of movements. For writers, maps were an invitation to travel in their imagination. In other words, geography became a compulsory reference, and it changed the nature of narrative literature.

As these remarks and the quotation from *Tristram Shandy* remind us, geography in the eighteenth century was still very far from the scholarly compartmentalized discipline we know today. Its only aim was to give a full account of the surface of the earth. To this must be added that it continued to evolve in the course of the century. The aspect that appealed to the public most directly was its mapping of the globe, still uncompleted at the end of the century, and the information it provided on its size, its salient physical features, the climate of the different regions, their populations and what we would today call their biodiversity. Geographers were first and foremost cartographers of the earth and the sky. It must not be forgotten that mapping included the study of astronomy, familiarity with the vault of heaven being as essential

8 Matthew H. Edney, "Reconsidering Enlightenment Geography and Map Making: Reconnaissance, Mapping, Archive", *Geography and Enlightenment*, 190. By 'public' was essentially meant the reading public of the new press. Literacy was more widespread in England than in many other European countries but far from being universal.

9 Hans Robert Jauss, *Pour une esthétique de la réception* (Paris: Gallimard, 1978), 49.

to navigation as to explorations on land. Since Christopher Columbus, Vasco de Gama and Magellan, progress in the knowledge of the globe had gone hand in hand with overseas voyages. Most useful were the sailors' practice of systematically plotting their positions at sea or when they landed on coasts. Equally indispensable for the mapmaker were the travellers' charts, graphs and other draughts. Recording and comparing the travellers' notes and sketches helped determine the location of places they sailed by or stopped at with increasing precision. Mathematics played a capital role in the elaboration of this reasoned description of the earth. As the first encyclopaedia, Ephraim Chambers's *Cyclopaedia*, explains, geography became "a branch of mathematics, of the mixed kind: in that it consider[ed] the earth and its affections, as depending on quantity, and consequently as measurable."[10] The painstaking compilation of the figures made it possible to get, progressively, a more accurate picture of the size and shape of continents, of the extent of the oceans and of the position of islands.

Mathematics also provided the framework necessary to produce handy visual representations of the earth's surface or the night sky in the form of charts, globes or hemispheres. Basically, the work consisted in choosing a suitable scale, in drawing the lines for latitudes and longitudes on globes and hemispheres, or the equidistant straight lines of meridians and perpendicular straight lines of parallels in Mercator projections, after which there remained to fill the grid—specialists prefer the term "graticule"— with the extent and shape of land masses and the location of particular spots that one wanted to situate on the earth's surface or a section of it. Because their information came from seamen, cartographers were particularly successful in drawing the picture of coastal areas. Theirs was a geography of shapes and contours. Globes, hemispheres and maps brought order to the mass of information. They "replace[d] reality with a more useful and simplified picture of commonly accepted salient aspects of reality".[11] They enabled people to take in the whole earth or large portions of it at a glance and to locate the position of any place referred to in everyday life. More important, as another historian adds, cartographic depictions served to

10 Ephraim Chambers, *Cyclopaedia* (London: 1728).

11 Anne Marie Claire Godlewska, "From Enlightenment Vision to Modern Science? Humboldt's Visual Thinking", *Geography and Enlightenment*, 250.

guarantee that an indissoluble link existed "between words, drawing and things".[12] Maps synthesized geographical knowledge by providing a condensed version of field observations.[13] In other words, they became the central feature of people's geographical "horizon of expectation". In everyday life, maps were increasingly used in the planning of people's movements. In both respects, cartography was bound at some point to attract the attention of authors of adventure stories. The determination of this moment was the starting point of the present study of geographical representations of space in eighteenth-century literature.

Progress in this charting of the surface of the globe had however been slow and it continued to be slow throughout the eighteenth century. Not only was the sifting of relevant information among a mass of data difficult but it was also difficult to get precise coordinates. This meant that, for all its claims to mathematical accuracy, mapping remained imprecise and sometimes bordered on fiction.[14] It is little wonder that, although unknown areas shrank and main landmasses slowly assumed their present shapes and positions, the mapping of the earth was still the target of criticism on the part of sceptics when *Robinson Crusoe* and *Gulliver's Travels* were written. Among the main difficulties facing mapmakers was the rudimentary character of the measuring instruments. With tools like compasses and sextants, measurements were at best approximate and, until the invention of Harrison's chronometer in 1760, it was for instance impossible to determine longitudes systematically. Navigators would mostly depend on such rare occurrences as eclipses of the moon or the satellites of Jupiter to calculate them. Another problem was that much geographical information was either kept private by dynasties of navigators or secret by governments in order to protect their interests against competitors. Thus Spanish navigators appear to have been more familiar with the Pacific than the Dutch and the English and the Dutch better acquainted with some of the coasts that were eventually identified as those of the

12 Antoine Picon, *French Architects and Engineers in the Age of Enlightenment*, Cambridge Studies in the History of Architecture (Cambridge: Cambridge University Press, 1992), 214.

13 Matthew H. Edney, "Reconsidering Enlightenment Geography and Map Making", 175.

14 On the lack of precision of eighteenth-century maps and cartography in general, see Henri Bouasse, *Géographie mathématique* (Paris: Delagrave, 1919), 287–91.

Australian continent.[15] In the naval war that European powers waged against each other, a prized item, when an enemy ship was seized, were the charts it had on board. In 1712, a sea captain added an appendix to an account of his circumnavigation of the globe. Its detailed description of the American coast roughly from Acapulco to the south of Chile was subtitled: "from the best Spanish Manuscripts taken in the South-Sea."[16] Not even the sea most familiar to Europeans had been properly mapped: the distance between Alger and Alexandrette, now Iskenderun (south-east Turkey), was reckoned to be nearly one third longer than it actually is. Last but not least, maps were seldom the work of qualified geographers. Mapping was a craft mostly left to engravers, who were as likely to feel as much concern for the aesthetic qualities of the picture as for its accuracy. The mapping of the globe only attained some final stability after James Cooke's expeditions in the Pacific between 1769 and 1776. At the end of the century, only the poles remained unchartered and, as every reader of Joseph Conrad's *Heart of Darkness* knows, the interior of Africa. What had changed was that the authority of maps was no longer questioned. The lack of precision of maps could, of course, be food for the imagination beside a target of criticism.

The other important aspect of world geography that interested the public apart from its cartographic representations was the collecting and collating of information on the various areas of the globe. This huge undertaking, which consisted in sifting data from navigators', merchants', adventurers' and soldiers' logbooks, reports and memoirs, was encouraged in England by The Royal Society of London for the Promotion of Natural Knowledge and, on the Continent, by similar bodies of scientists called academies and patronized by their respective sovereigns. The idea was to constitute a corpus of geographical knowledge about the populations and the resources of the world. The method was strictly empirical and descriptive.[17] It aimed at encyclopaedic comprehensiveness. Atlases became the depositaries of this information

15 On the case of Australia, see "Mapping of New Holland", an interesting note on the NET site <http://www.nla.gov.au/exhibitions/southland/Mapping-of_New_Holland.html> of the National Library of Australia, 1–2.

16 Woodes Rogers, *A Cruising Voyage round the World* (London: 1712).

17 Margarita Bowen, *Empiricism and Geographical Thought: From Francis Bacon to Alexander von Humboldt* (Cambridge: Cambridge University Press, 1981).

and played a central role in the diffusion of knowledge. Through the flow of factual observation, Europeans were able to build up a more precise picture not only of overseas or continental routes but also of life in remote areas of the earth. Text and maps were indissociable. Together they created a discourse "replicat[ing] the world as a conceptual museum that could be indefinitely expanded to accommodate ever more extensive and intensive geographical information."[18] This body of knowledge classified by regions or countries, this archive of descriptive information as it is often described, contained treasures for the curious reader.

Meanwhile, the advent of the modern state called for the mapping of individual countries and regions, in other words for new types of geographical representations susceptible to extend the government's control over the country. Here, the problem was no longer to locate places but to get a more precise picture of land surfaces, borders and the division into territories, as well as of the distribution of the population, of the economic activities and of natural resources than had been available so far. Special attention was paid to the improvement of communication routes. Even though the groundwork for small- and large-scale "plans" was not, like world maps, left to amateur geographers, but was entrusted to professionals, land surveyors, civil and military engineers, the progress of cartography at the level of the kingdom or county was equally slow. The determination of geographical coordinates by astronomical observation and systematic triangulation were difficult to apply on a local, regional and national scale, and when they were, the results were sometimes unexpected. Thus, in 1682, Jean-Félix Picard and Philippe de la Hire published a first tentative map of France based on triangulation, *Carte de France, corrigée par ordre du Roi sur les observations de Mrs de l'Académie des sciences*. It showed that, like the Mediterranean, Louis XIV's kingdom was smaller than formerly believed, which apparently led the monarch humorously to remark that "the gentlemen of the Académie des sciences had made him lose more territory than his enemies."[19] In England, the survey was not completed until the end of the eighteenth century. This meant that the calculation

18 Matthew H. Edney, "Reconsidering Enlightenment Geography and Map Making", 190.

19 "les messieurs de l'Académie lui ont fait perdre plus de territoire que tous ses ennemis." <fr.wikipedia.org/wiki/Histoire_de_la_triangulation_en_France>.

of distances remained approximate and people preferred to speak of the number of days a journey took.

During the second half of the century, increasing attention was also paid to topography, the relief of the land, for whose representation cartography was ill adapted. For the preparation of military operations, topographers had never ceased to resort to chorography, namely the drawing of series of panoramic pictures of an area, of landscapes, from different viewpoints usually at ground level to present theatres of operations.[20] In the eighteenth century, many did so with artistic skill and exerted their talent outside the military domain. In their different manners, the representations of the national space and its regions reinforced people's consciousness of belonging to an area, to an economic network and to a country. It also sharpened their awareness of their proximity or distance from the centres of power. As geography embraced new fields of investigation, still other more specialized types of maps or graphs soon emerged focusing on particular aspects of the physical, vegetable or human world.

That mapping came to play such a prominent role in the thinking of the period that it became a byword for the organization of encyclopaedic knowledge was not simply due to the interest people took in the globe and their or other countries. Every physical space is not geographical and mapping extended to many other activities involving the concrete organization of space. Cartography was only the most prestigious in the eyes of the public. Besides the planning of new urban extensions, of which Bath was an outstanding example, there were the building of spacious houses, the shaping or reshaping of all sorts of public and private spaces from botanic gardens and entertainment parks to simple orchards and private grounds,[21] the development of systems of canals and the designing of new types of fortifications. For all these, land surveyors and architects drew plans according to the same principles of mathematics and geometry as geographers. While

20 M. K. Barrit (Captain Royal Navy), "John Thomas Serres and the British Blockade, 1799–1800", Papers of the First International Paul-Gabriel Boucé Conference, Paris, 2006, due to appear in *The Sea: A Main Motif for an Aquatic Enlightenment* (Brooklyn, NY: AMS Press, 2013).

21 Simon Varey, *Space and the Eighteenth-Century Novel* (Cambridge: Cambridge University Press, 1990). This study begins with a discussion of the discourses of architects and focuses on the use of available space.

architects, writers and thinkers were showing increasing interest in the spaces of human activities, European kingdoms continued to expand their colonies and overseas markets. From individual elbow room to the extension of property and colonial expansion, the control and organisation of physical space became a central concern of the period. A link may finally be established with two apparently unrelated domains: Locke's empiricism and Newton's mechanics. In Locke, the conception of space as existent reality was central to explain the acquisition of knowledge through the senses, and in Newton, space was a prerequisite for the measurement of the size and motion of bodies in the universe.[22] Between the late seventeenth and early nineteenth centuries, the measuring of the environment and its reshaping were at the centre of a revolution in the way people saw the world. Spatialization was the word in every field. In their analyses of the political, social, economic and cultural problems of the period in terms of spheres and realms, philosophers and literary historians from Jürgen Habermas to Michael McKeon underline this and demonstrate that these spatial metaphors were the very ones that people used from the seventeenth century onward to re-think their presence to the world and to describe the changing patterns of public versus private life and personal relations.[23] Being omnipresent, representations of space could not but make people increasingly aware of occupying an environment and of the possibility of moulding it to their needs or ambitions. To deal with this second aspect of eighteenth-century representations of space, I have used, rather freely, Henri Lefebvre's notion of "production of space".[24] As a critical tool it has proved helpful, shedding light on the new dimensions of experience opened up by geography. The different chapters of the present study discuss a sample of these "productions of space". In a context where sight took precedence over the other senses and graphs over text, geography offered writers a rich and supple instrument to think anew about life and fictionalize it.

22 Ivor Leclerc, "Concepts of Space" and Hopewell Selby, "'Never Finding Full Repast': Satire and Self-Extension in the Early Eighteenth Century", *Probability, Time, and Space in Eighteenth-Century Literature*, ed. Paula R. Backscheider (New York: AMS Press, 1979), 209–16, 217–47.

23 Jürgen Habermas, *The Structural Transformation of the Public Sphere* (Cambridge, Mass: MIT Press, [1962] 1989); Michael McKeon, *The Secret History of Domesticity* (Baltimore: Johns Hopkins University Press, 2005).

24 Henri Lefebvre, *La Production de l'espace* (Paris: Anthropos, 1974).

At the risk of oversimplifying a very complex evolution, one could say that this way of conceiving space in geographical and geometrical terms was one more blow to what remained of the old mediaeval view of the universe and to the belief in a stable hierarchical world-order binding and ranking God's creatures, from the pure angelic spirits to man endowed with life and understanding and to inanimate objects endowed with mere existence. It was commonly pictured as "the Great Chain of Being", a vertical metaphor emphasizing the subordination of each rank to those above it and its dependence on the others.[25] Two secular versions of this belief still ruled the relations between a monarch and his subjects and between a father and his family. They had hardened into a social dogma but were being slowly eroded. Increasingly aware of occupying a physical, social and economic place and space of their own in their world, individuals adopted a spatial outlook and this gave them a strong sense of their individuality.[26] Most of the fictional works studied hereafter feature characters with individualist bents. Is not the first of them, Robinson Crusoe, a son that rebels against the plans that his father had for his future? This was symptomatic.

It is easy to see why writers adopted a more geographical, or at least spatial, vision of their surroundings and to understand the new turn this gave to their representations of life. On the other hand, it is unexpected to find that the representations of space in geography and literature underwent a parallel evolution, considering that their discourses were drifting apart, one becoming increasingly visual and diagrammatic and the other remaining verbal. Yet this is what happened. In both geography and literature, there was a progressive shift from plane cartographic representations of space to three-dimensional chorographic representations, as well as a diversification of these representations to respond to new domains of interest. It has been possible to identify four moments in the evolution of literary representations of space, corresponding to four successive orientations: explorations, road travels, Pre-Romantic rambles and Romantic wanderings. A further surprise was that each new

25 Arthur O. Lovejoy, *The Great Chain of Being* (Cambridge, Mass.: MIT Press, 1936). E. M. W. Tillyard, *The Elizabethan World Picture* (Harmondsworth: Penguin, [1943] 1981), 33.

26 John Richetti, *Popular Fiction before Richardson: Narrative Patterns, 1700–1739* (Oxford: Clarendon, 1969), 35, 62–67.

literary representation of geographical space simultaneously highlighted a different way of moving through space.

However flagrant their rapprochement, the geographical and literary discourses on space naturally pursued different goals, as they still do, and did not appeal to the imagination in quite the same way. Geography conceptualized space and aimed at offering a comprehensive vision of the surface of the globe or precise pictures of sections of it. Literary works used the "simplified pictures" of geography to describe the way individuals live space. Writers did not hesitate to mix exactitude and invention, which the geographical discourse never did, at least intentionally. In eighteenth-century literary representations of space, the amount of geographical material also varied greatly from work to work. They ranged from the mere mention of place-names, the writer relying on the readers' horizon of discourses to fill in the background, to complete descriptions of landscapes. Whichever the procedure, it posed no problem to the authors' first readers, who shared his geographical "horizon of expectation". Things are different for later readers. Places change though their names may not. The representations of geographical space in the literary works of the past in general, and of the eighteenth century in particular, raise with special acuity the question of what is left unsaid by a writer or merely hinted by him. Reading works of the past always implies a leap into the more or less dimly known and even though the imperfectly known is part and parcel of a fundamental dimension of life, to miss its existence can only impoverish, if not partly invalidate, our understanding of a work. Owing to their reliance on the common knowledge of a past age, geographical representations are a particularly sensitive case. Recourse to the collective archive that made it unnecessary for the author to spell things out for his early readers can prove useful.

As already suggested, geography is not concerned with the distinction between private and open or public spaces. The point has to be made as, in recent years, a great deal of attention has gone to the study of privacy, a dimension of life that became increasingly important in the eighteenth century and was also conceived and lived in spatial terms. What follows looks in the opposite direction: it ignores private apartments and closets, these typical spaces of privacy, but it embraces, on a par with villages or islands, the location of the houses people inhabited and privately owned lands. Although the distinction public versus private is not relevant to geography, it had a consequence that was related to the present subject: it made the public sphere of life look to people more resolutely public, as will be seen in several works examined hereafter.

References to a geographical space in literature are seldom studied for themselves. They are normally seen as a subordinate aspect of creative writing, a backdrop, except when they are used to enhance the symbolic and thematic patterns that make a work unique. They tend to be viewed as unoriginal and of little significance beyond lending a frame or background to an action. At first sight, nothing could be less interesting than the descriptions of the itineraries followed by two travellers to Paris or hikers in the Lake District. And yet these apparent or real similarities often matter. They play variations on a same tune. As already said, as references to a "horizon of expectation" common to writer and reader, they anchor the representations in their times by linking them to contemporary contexts, which the historian of literature cannot neglect.

As intimated by the epigraphs, the present examination of the changing representations of geographical space in the long eighteenth century takes its stand at the intersection of history and the phenomenology of perception. The nature of the literary works under consideration varies, sometimes considerably, so that they can in no way be said to form a corpus. In fact, their treatment of topography is often the only common denominator between them. This explains why the present study focuses on this topographical character of their representations of space and not on the aesthetic qualities of the latter. The focus on the representation of topography has naturally dictated the choice of the works discussed here as well as the procedure of the study by soundings. This is in no way intended to suggest that other literary works by the same or other writers might not have served the same purpose. That still other works of the same period did not exhibit this interest in geography is of little relevance to the present project, which is to define a trend and assess its importance.

As a recent essay puts it, geographical and literary discourses are both the expression of "human beings [who] are part of the earth, the

27 Ihab Hassan, "Maps and Stories: A Brief Meditation", *The Georgian Review* 59: 763 (2005).

sky, the sea, and [whose] maps are part of the mind, mind seeking its answering part in the world."[27] However much their modes of expression came to diverge, they shared the same plots and the same poetics. Cartography with its accompanying descriptions of the different areas of the world and literature explored the same subject matter: movements, sea routes, roads, inns, encounters and adventures. They aroused the same reactions in readers: feelings of wonder or vertigo before vast spaces and strange elsewhere's, of awe in front of the infinite variety of nature. They awakened a lure of blessed islands, the thrill of sea passages and untrodden paths. Perusing maps, atlases, guidebooks, genuine or fake travel accounts, the public enjoyed the twin discourses much in the same way. In literature, the representation of geographical space can even be said to have diverted to itself some of the attention that had previously gone exclusively to the adventure, or to the fable and ethical message of the works, and it is not totally impossible that it was this diversion of attention towards the diversity of the world that Swift, the satirist and moralist, had in mind when he made Gulliver suffer four times the same disillusionment with his travels. But his protest was a rearguard action. With Defoe and genuine travel accounts, the magic of the world had carried the day. For inhabitants of the earth today, the same magic still works, however much the world has changed. The leap into the past adds an extra imaginary dimension to geographical representations, but the basic geography of the globe remains an easy and attractive point of entry into the works of eighteenth-century writers who responded to the attractions of geography, as well as into the history of these works. Our different perceptions, our outlooks remind us that if the surface of the globe is constantly changing it is partly at least due to the human impact on it. To look at the representations of what this surface once looked like, and at how people experienced it, heightens our awareness of our historical and cultural moment and of our distance from the works we read, and this enriches our experience of it and of the world we live in.

1. The Sea in Defoe's and Swift's Fictions[1]

Navigation played a prominent role in the development of modern geography and it is fitting that this study should begin with eighteenth-century literary representations of maritime spaces. Here two famous works of fiction immediately come to mind for their conspicuous representation of world geography and its seas: *The Life and Strange Surprising Adventures of Robinson Crusoe, of York, Mariner* (1719) and *Travels into Several Remote Nations of the World, in four parts, by Lemuel Gulliver* (1726). Swift's *Travels* and Defoe's *Robinson Crusoe*, whose fourth edition was published five years earlier, were also the first works of English fiction to include maps and make use in their narratives of "realistic" maritime geography, "realistic" according to the standards of the time.[2] The cultural context easily accounts for the idea of using maps in fiction. It is well known that both fictions were inspired by the abundant travel literature available at the time, some of which, quite recent, did include maps. This was in particular the case of *A New Voyage round the World* (1697) by the adventurer, William Dampier. His travel account set a pattern for other reports by William Funnel, Woodes Rogers and Edward Cooke.[3] The last two were captains with whom Dampier also sailed. Then, there was the spreading popularity of maps and atlases.[4] Swift and Defoe were the contemporaries of several

1 An early version of this chapter was presented at the First International Paul-Gabriel Boucé Conference, Paris, 2006 and is due to appear in *The Sea: A Main Motif for an Aquatic Enlightenment* (Brooklyn, NY: AMS Press, 2013).

2 Sir Thomas More had anchored his Utopia to the world by means of a map but, owing to the nature of his Nowhere, the geography was purely imaginary.

3 William Funnel, *A Voyage round the World* (London: 1707); Woodes Rogers, *A Cruising Voyage round the World* (London: 1712); Edward Cooke, *A Voyage to the South Sea and round the World Performed in the Years 1708, 1709, 1710 and 1711* (London: 1712).

4 On English cartography and its popularity, see Helen M. Wallis, "Geographie is Better than Divinitie: Maps, Globes, and Geography in the Days of Samuel Pepys" and David A. Woodward, "English Cartography, 1650–1750: A Summary", in *The Compleat Plattmaker: Essays on Chart, Map, and Globe Making in England in the Seventeenth and Eighteenth Centuries*, ed. Norman J.W. Thrower (Berkeley: University of California Press, 1978), 1–43, 159–193.

active cartographers, two of whom they either explicitly or implicitly refer to in their works, Herman Moll (1654–1732),[5] and John Senex (1690–1740), whose maps opened the doors of the Royal Society to him. John Senex is also mentioned as one of the printers, in other words one of the publishers, of Defoe's world atlas of 1728.[6] The map in the fourth edition of *Robinson Crusoe* and the geographical setting in Defoe's novels have been more often taken for granted than discussed. As for Swift's maps and geographical information in *Gulliver's Travels*, they have attracted some attention, but discussion of this aspect of the satire has usually been limited to the question of their authorship and accuracy, or rather its lack.[7] History and semiotics suggest other ways of looking at them, and what follows is intended to draw attention to the significance of their faithful representations of the globe or combination of accuracy and fantasy, and to the iconic character of the maps. Limited to *Robinson Crusoe* and *Gulliver's Travels*, the use of maps has at least the merit of drawing attention to the nature of the two authors' depictions of geographical space in their fictions and of highlighting two widely different ways of understanding space and visions of the world.

Defoe's interest in geography was a life-long affair. Geographical concerns are omnipresent in his work. They find their clearest expression in three late works, all three compendiums of his vast knowledge of the subject: a descriptive guidebook, *A Tour through the Whole Island of Great Britain* (1724–1726), an atlas, *Atlas Maritimus & Commercialis;*

5 Burton J. Fishman, "Defoe, Herman Moll, and the Geography of South America", *Huntington Library Quarterly* 36 (1973): 227–38. Swift refers to Moll in the *Travels* (see below). In *The Cartographer and the Literati: Herman Moll and his Intellectual Circle* (Lewiston: Edwin Mellen Press, 1999), Dennis Reinhartz argues that Dampier and Defoe were acquaintances of the cartographer, but the case he makes for Swift is not so convincing.

6 See below.

7 John Robert Moore, "The Geography of *Gulliver's Travels*", *The Journal of English and Germanic Philology* 40 (1941): 214–28; Frederick Bracher, "The Maps in *Gulliver's Travels*", *The Huntington Library Quarterly* 8 (1944): 59–74; Arthur E. Case, "The Geography and Chronology of *Gulliver's Travels*", in *Four Essays on "Gulliver's Travels"* (Princeton: Princeton University Press, 1945), 50–68. The most recent examination of Gulliver's geography is Dirk F. Passmann's *"Full of Improbable Lies": "Gulliver's Travels" und die Reiseliteratur vor 1726* (Frankfurt/Main: Peter Lang, 1987), 235–74.

or, a General View of the World, so far as it relates to Trade and Navigation (1728), which includes detailed descriptions of continents, islands and their resources, and *A General History of Discoveries* (1725–1726), a work to set beside Herman Moll's own historical atlases.[8] In his fiction, Defoe relied heavily on geography to create an illusion of authenticity and render the memoirs of his main characters credible. The world map used in the fourth edition of *Robinson Crusoe* is a graphic confirmation of this. To describe the travels of his heroes, he borrowed from the contemporary geographical discourse.[9] He did it so literally that, on the whole, the procedure can be termed transparent, which accounts for the little attention given to it. But the resulting handling of space in the novels is not less interesting for all that.

The sea holds a prominent place in Defoe's work. His novels evidence a predilection for the very seas and oceans sailed across by the authors of the genuine memoirs referred to above, in particular William Dampier: the North and South Atlantic in *Robinson Crusoe* (1719), *Moll Flanders* (1722) and *Colonel Jacque* (1722); the Caribbean and the Gulf of Mexico again in *Robinson Crusoe*, *Captain Singleton* (1720) and *Colonel Jacque*; the Indian Ocean and China Seas in *The Father Adventures of Robinson Crusoe* (1720) and *Captain Singleton*. To these, his *A New Voyage round the World* (1725), whose title is a deliberate borrowing from Dampier's or Rogers' memoirs and whose publication precedes *Gulliver's Travels* by two years, adds an exploration of the South Seas, as the South Pacific was then called. It is a supposed relation inspired by a voyage of two Dutch navigators a century earlier, Willem Schouten and Jakob Le Maire.[10] The Mediterranean plays an insignificant role in Defoe's work, which reminds us that, while the economic centre of the world had moved to the west, from the Continent to England, the interest of readers had shifted even further east and west, to the Indian Ocean, the Far East, the West Indies and the Pacific.

8 In particular, Moll's *Geographia Antiqua Latinorum et Graecorum* (London: 1721).

9 On Defoe's sources of geographical information, see Arthur W. Secord, *Studies in the Narrative Method of Defoe* (Urbana: University of Illinois, 1924), 32–38, 49–62, 68–74, 89–96, 109, 118–19, 129–38, 146, 147–60, 236, 241–43.

10 *A New Voyage round the World* (1725), 98. Further references to John McVeagh's edition (Pickering and Chatto) are given in the text.

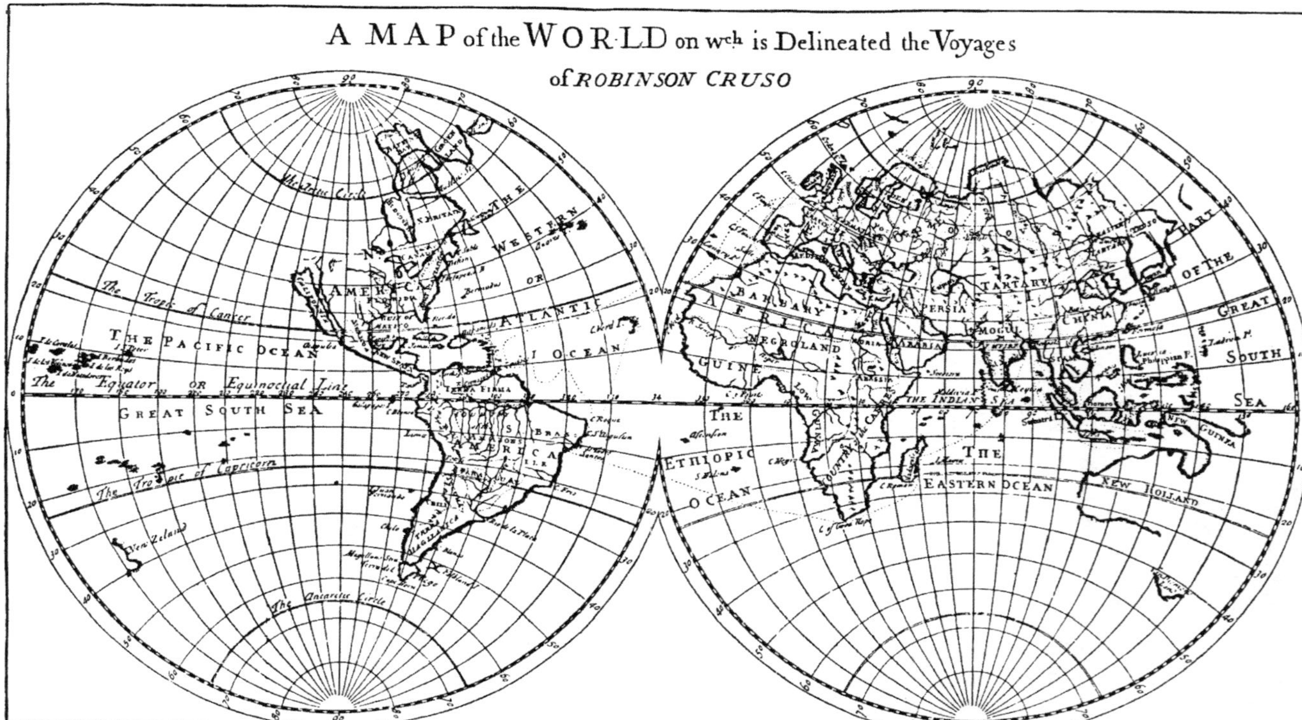

Fig. 1.1: Robinson Crusoe's travels. The dotted lines show both the voyage leading to the shipwreck and return to England and the later voyage to the Far East and return through Continental China and Russia.

The surface of the sea has always looked flat to the human eye, and it is little wonder that its geographical representations have usually been two-dimensional. Following the method of sea logbooks with their ships' logs and sailing directions, Defoe's narrative discourse suitably lays emphasis on his characters' sea routes. It is in connection with the diagrammatic sketching out of these routes that the flat cartographic nature of his fictional geography comes out most clearly. His *New Voyage round the World* gives a striking illustration of this with its story of a circumnavigation of the world. The captain, who is the anonymous author of the alleged memoir, narrates how his expedition sails from the Thames first to Nieuport in Flanders, then to Galway in Ireland before stopping at the Canary Islands. Much against his will, on leaving the Canary Islands, the voyagers take the same westward direction as all the previous circumnavigations: they cross the Atlantic to Cape St. Augustine in Brazil. From there, they follow the coast, passing the Bay of All Saints, today Todos os Santos,[11] Buenos Aires, down to Porto San Julian, where many navigators since Magellan had made a necessary stop before attempting the difficult passage to the South Seas. Two providential tempests, however, prevent them from rounding Cape Horn and send the ships from Porto San Julian to the Cape of Good Hope, which puts them back on the captain's intended track. From there, they follow the east coast of the African continent up to the southwest point of Madagascar. From the north of Madagascar, the vessels make for India, the Malabar coast, Ceylon (now Sri Lanka), the coast of Coromandel to reach Achin, a region of northeast Sumatra. They sail past the Straits of Malacca and Singapore, follow the northern coast of Borneo and reach Manila seventeen months and two days after their departure from England. After Manila, the expedition calls at Luzon or Luconia, the largest island in the Philippines and at the Ladrones. From the Ladrones, it follows a south-southeast trajectory down to the 50th meridian in the hope of discovering the Terra Australis Incognita of old maps, a southern continent that, geographers postulated, acted as a counterweight to the large Eurasian continent in the northern hemisphere.[12] Not finding it,

11 This is also the bay where Crusoe lands after his rescue off the African coast by a Portuguese merchantman.

12 Edmund Halley's 1700 expedition mentioned in the introduction had reached a similar southern latitude in the Atlantic. In the Pacific, James Cook was the first to reach the ice fields of the Antarctic fifty years after Defoe's fiction.

the vessels bend their course north-west towards the South American continent, cast anchor at the island of Juan Fernandez, made famous by Alexander Selkirk, and finally arrive at Valdivia, which Defoe spells Baldivia, as British cartographers usually did. From Valdivia, the expedition sails up the coast of Chile and Peru, calling at the ports of Conception, Santiago, Coquimbo, Copiaco, Arica and the port of Lima.[13] Back at Valdivia, the leader of the expedition decides to send a group of explorers from there across the Andes. Their mission is to find a route across the mountains, known to be rich in gold, to the Atlantic coast, while the vessels round Cape Horn, stop once more at Porto San Julian, then at Porto Deseado and Capo Blanco. At an unidentified place that Defoe calls the mouth of the Camerones River, the expedition re-embarks the party of scouts that have found a way across Patagonia. To get back to Europe, the expedition sails along the Atlantic coasts of both Americas up to Newfoundland to take advantage of the currents and winds. From there, it makes for the area of Bordeaux before reaching Dunkirk, its final destination, three months after leaving Argentina. The geographical background of the circumnavigation unfolds like the plane representations of the world on one of Moll's double hemispheres or Mercator projections.

Defoe's discourse of sea routes is a direct expression of the contemporary conception of "utilitarian geography" as a functional tool to get from point A to point B.[14] *Robinson Crusoe*[15] again illustrates this, albeit in a less obvious manner. The story of the hero's life begins with his departure from York, where his family live, describes the circumstances in which he sails from Hull but is shipwrecked near Yarmouth on his way to London (60–66), embarks on a vessel bound for "the coast of Guinea" (68), comes back to London, embarks a second time for the same destination, is made prisoner by pirates on the African coast, somewhere in the latitude of the Canary Islands (69), becomes a slave at Salé near Rabat in Morocco, escapes in a fishing boat, sails along the Atlantic coast

13 Defoe seems to have used information on this coast that he found in William Funnel's *A Voyage round the World*, chapters 6–8.

14 Paul Claval, *La Pensée géographique: Introduction à son histoire* (Paris: SEDES, 1972), 13.

15 *The Life and Strange Surprising Adventures of Robinson Crusoe (1719)*, ed. W.R. Owens, *The Novels of Daniel Defoe* (London: Pickering and Chatto, 2008). The references to *Robinson Crusoe* are given in the text.

of what is today Morocco and Mauritania, is rescued by a Portuguese ship a little further south from the place where he had been the victim of pirates (81), sails in it to Brazil (81), where he settles as a planter, until the fatal day when, on his way to Guinea to purchase slaves for the planters, his ship is driven off course by two successive tempests and he is shipwrecked on the famous uninhabited island (87–90). In *Father Adventures*, after leaving his island for the last time, Crusoe follows the trade route to the East Indies, round the Cape of Good Hope, past Madagascar and the coast of Coromandel, until he is put ashore in Bengal by a rebellious crew.[16]

Except when tempests blow their ships off course or when exploration is intended, Defoe's voyagers follow plied routes or, at least, routes that would become so, because, in his fiction, his "utilitarian geography" is already the commercial geography of his future atlas. *Moll Flanders* and *Colonel Jacque* confirm this. Their main characters cross the Atlantic to Chesapeake Bay and back several times. They do so by the usual route of British ships, which is also that of the transportation of criminals and kidnapped people used as cheap labour force in Virginia and Maryland. All three novels mention the European products taking the same routes to meet the needs of planters in the colonies while, in the opposite direction, colonial produce, and money in the form of letters of exchange are shipped to England. Defoe never forgets to detail such things. Crusoe has "all Sorts of Tools, Iron-Work, and Utensils necessary for [his] Plantation" sent to him by a merchant in London. He also receives "*English* Manufactures, such as Cloath, Stuffs, Bays", which he manages to sell to advantage (84). The tools and manufactures travel first to Lisbon by a normal sea link between allied countries, and from Lisbon to Brazil by the regular trade route between Portugal and its colony. As one might expect, smugglers take the same routes and pirates thrive on them. In his three voyages from Virginia to the Gulf of Mexico to supply Spanish merchants at Veracruz with illegal British goods, Colonel Jacque must follow the same routes as the Spanish fleet.[17] As for Captain Singleton and his

16 *The Father Adventures (1719)*, ed. W.R. Owens, *The Novels of Daniel Defoe* (London: Pickering and Chatto, 2008), 119–41.

17 *The History and Remarkable Life of the Truly Honourable Col. Jacque*, ed. Maurice Hindle, *The Novels of Daniel Defoe* (London: Pickering and Chatto, 2008), 249, 252, 253, 256.

fellow pirates, after attacking the local traffic in the West Indies, they move to another hunting territory in the East Indies, and there give chase to merchant ships—their purchases as their quarries were called in the language of piracy—on the spice and Vietnam routes (123–27, 156–70).[18]

A New Voyage round the World presents Defoe's most ambitious handling of commercial geography. The circumnavigation of the globe proceeds, the subtitle says, "by a course never sailed before". As seen above, after the false start westwards, the voyage finally reverses the direction of all previous navigators since Magellan, and it does so precisely to turn it into a profitable trade route. The novel is the story of an enterprising sea captain, who sets out to get round Spain's protectionist policy in its colonies and to open a direct route for English manufactures first to the Philippines and then to Chile and Peru. The sea captain and head of the expedition, a former privateer turned promoter of "open Commerce," (96) does so by a method of striking modernity. His venture is financed by British merchants, who intend to set up "an East-India Company in Flanders" according to the title, in other words, to create an international company with its seat on neutral ground. The Captain also takes on an international crew to facilitate contacts with Spanish markets. The expedition practises a transport of goods without transhipment to undersell Spanish and French competitors and uses the modus operandi of what is called logistics today, namely the purchase of local products on the way, part of which can be sold at the next markets so as to reduce the costs of a long navigation, while the rest of the cargo of exotic wares is brought back to England (41–42, 89–91, 152, 156).[19]

Defoe's utilitarian and commercial geography in his fiction is so precise that it does not need the support of maps. In fact, his narrative discourse is itself sufficiently cartographic to dispense with them. In this sense, it is not surprising that the author did not, as a rule, preface his narratives with them and that the addition of a map of the world to the fourth edition of *Robinson Crusoe* was a late idea. There, as in Dampier,

18 *The Life, Adventures, and Pyracies, of the Famous Captain Singleton (1720)*, ed. Maurice Hindle, *The Novels of Daniel Defoe* (London: Pickering and Chatto, 2008). The references to *Captain Singleton* are given in the text. To purchase, from the French *pourchasser*, to hunt down.

19 For more details, see Jean-Paul Forster, "*A New Voyage round the World* de Defoe: le globe, théâtre nouveau d'activité", *Territoires, RSÉEA XVII–XVIII* 66 (2009): 189–201.

the double hemisphere simply prefaces the narrative with an icon of the scope of Crusoe's adventures (fig. 1.1). With its cartographic depiction of sea routes across the globe, on which are woven various stories of intercontinental travels and exchanges, Defoe's fiction, like his world atlas, offers a geographical representation of the world on an unprecedented scale. It also depicts a world in movement, with planters, slave dealers and slaves, forced and freelance labourers crossing oceans on merchant ships, navies cruising along coasts, European and Far-Eastern goods being exchanged and exotic products and wealth in the form of letters of exchange pouring into England. As social environment, his seas are no different from land: they are navigated by honest and hardworking people, not-so-honest people, and criminals. In this maritime world, buccaneers sometimes seem to be a greater danger to navigation than tempests. Yet, the general note is optimistic. Failures are followed by achievements and criminals repent, at least in so far as the main characters are concerned. Politics intervenes little in the picture, with the exception of references to the relations between England and Spain, but even then, *Colonel Jacque* and *A New Voyage* show that the problem is more economic than political, a question of bypassing protectionist barriers. The novels all agree on one point: it is easier for merchants, sellers and buyers alike, to come to a good understanding than for governments. The promise of doing good business is sufficient for the development of friendly contacts. Defoe was a free trader before the term had been coined. As his *History of Discoveries* idealistically puts it, "Trade knows no Parties, no Politic, no religious Interests: Commerce is a certain Communication of Nations occasion'd by the necessities and for the good of Mankind; the Enemies of Trade are Enemies of all Men."[20]

The geographical representation of the maritime world in *Gulliver's Travels* is basically similar to Defoe's in his novels, a thing seldom acknowledged. A great deal of the information contained in the first and last chapters of each of the four "Voyages" consists of genuine contemporary geography in every respect as transparent as Defoe's, while the rest relates to the islands and lands supposedly discovered by Gulliver.

20 *A General History of Discoveries and Improvements (1725–26), Writings on Travel, Discovery and History by Daniel Defoe*, vol. 4, ed. P. N. Furbank (London: Pickering and Chatto, 2001), 118.

As all previous studies on the subject rightly conclude that it is impossible to situate the hero's discoveries without emendations of either the maps or the text or both, it has seemed more productive to start by sidestepping this specific issue, which obviously leads nowhere, and to see first what light the recourse to genuine geography sheds on the satire. Considered without its imaginary aspects, the geography of the *Travels* proves to be not only precise according to the state of knowledge of the time but also, like Defoe's, largely two-dimensional and cartographic in nature, utilitarian and economic rather than commercial: an authenticated geography of busy seaports, ports of call and the destinations of merchant ships. It draws on the same sources, which would have been familiar to readers of atlases and gazettes and, for certain details, to patrons of coffee houses, so that, although not foregrounded, it would have immediately held the attention of early readers. During the sixteen years of his nearly constant absence from home, Gulliver's family live first at Wapping (20) and then at Rotherhithe, or Redriff as Gulliver chooses to call it (80).[21] These two places were districts of docks respectively on the north and south sides of the Thames, about two miles down from what is today Tower Bridge, where people whose calling was connected with the sea resided. Among them, there were, beside shipmasters, pilots and instrument makers, sellers of hand-drawn portulan sea charts and printed maps, a fact which cannot have escaped Swift.[22] In Book I, Gulliver sails from Bristol, the most important port after London (20) and in Book IV, from Portsmouth (221). In the third voyage, he sails from the Downs (83), "a roadstead off Sandwich and Deal on the coast of Kent",[23] where ships used to drop anchor on their way to or from London Harbour, either waiting for a favourable wind and tide or for last dispatches before setting sail. It is always at the Downs that he lands on his return to England (79, 148, 218, 289), as William Dampier had done on his return from his first circumnavigation

21 *Gulliver's Travels*, ed. Herbert Davis (Oxford: Blackwell, 1965). The references to this edition are given in the text.

22 Thomas R. Smith, "Manuscript and Printed Sea Charts" and Coolie Verner, "John Seller and the Chart Trade in Seventeenth Century England" in *The Compleat Plattmaker*, 50–51, 91–92, 135.

23 *Gulliver's Travels*, ed. Angus Ross (London: Longman, 1972), 306.

of the globe.[24] In Books III and IV, Gulliver comes back to England by way of Amsterdam (217) and Lisbon (288). Amsterdam had lost its position as the economic capital of Europe to London but it was still a busy seaport, and Lisbon was the capital of the country that had opened the sea route to the Far East two centuries earlier. Two ports of call are also mentioned in the *Travels*. In the second voyage, Gulliver's ship has to winter at the Cape (83) and, in the fourth voyage, he stops briefly at Tenerife in the Canary Islands (221). Most of these places are also mentioned in Defoe's novels.

Like Crusoe, Gulliver sails on merchant ships in search of profitable trade. The routes followed and the intended destinations of his ships in "the East and West-Indies" (20) or the South Seas, namely the Pacific, were all familiar to well-informed contemporaries (fig. 1.2).[25] In the first three voyages, Gulliver sails first along the west coast of the African continent down to the Cape of Good Hope as hundreds of ships had done since Vasco de Gama had opened the way two hundred years before. From there, they either travelled across the Indian Ocean or up the east coast of the continent to Madagascar and to India or even further through the Indonesian Archipelago to North Vietnam. The destinations listed are "The South Sea" in Book I (20) and, more specifically in the other voyages, Surat (Book II, 83), Fort St. George, which was to become Madras and is now Chennai,[26] and Tonkin—now North Vietnam— (Book III, 154). Of these destinations could be said what has been said of *Captain Singleton*, that their mention "reflected … a world in which British trading posts were already established in India".[27] For Dutch ships, a current destination was Nagasaki in Japan (Book III, 217).

24 William Dampier, *A New Voyage round the World* (London: Argonaut Press, [1697] 1927), 367.

25 Herman Moll's map of 1697 has been preferred here to that of 1719, which is said to have been the basis of the regional maps prefixed to each voyage. See *The Correspondence of Jonathan Swift*, vol. 2, ed. David Woolley (Frankfurt/Main: Peter Lang, 2001), 373, footnote 9. The 1697 map is the one found in Dampier's own *New Voyage round the World*. Its two hemispheres show the then unknown areas of the world more clearly than the 1719 map and lend themselves better to the illustration of the trajectories of Gulliver's intended voyages.

26 Surat was a British trading post established in 1640, Fort St. George in 1639–1640, Bombay in 1665 and Calcutta in 1690.

27 Douglas Chambers, *The Reinvention of the World: English Writing 1650–1750* (London: Arnold, 1996), 24.

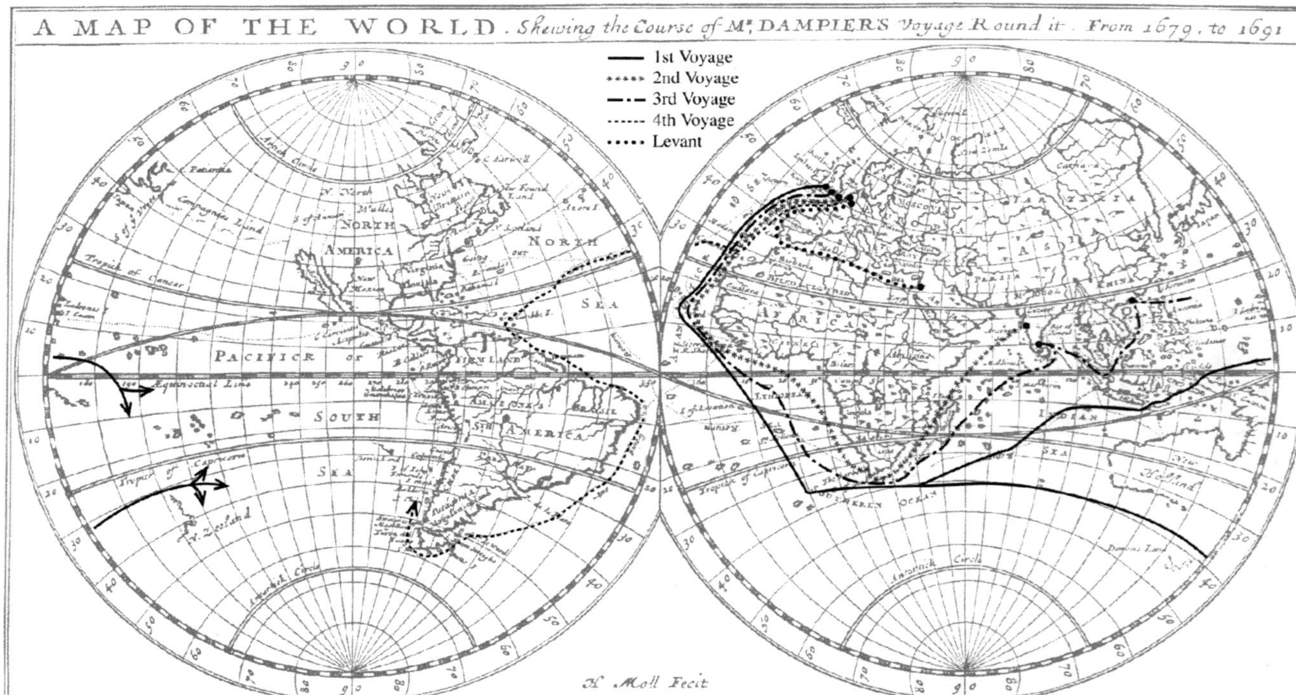

Fig. 1.2: The routes to Gulliver's intended destinations: 1st voyage: from Bristol to the South Seas, two possible routes; 2nd voyage: from the Downs, destination Surat; 3rd voyage: destination Fort St. George (Madras) and Tonquin (Vietnam); 4th voyage: to the South Seas via the Caribbean and South America; The route to the Levant.

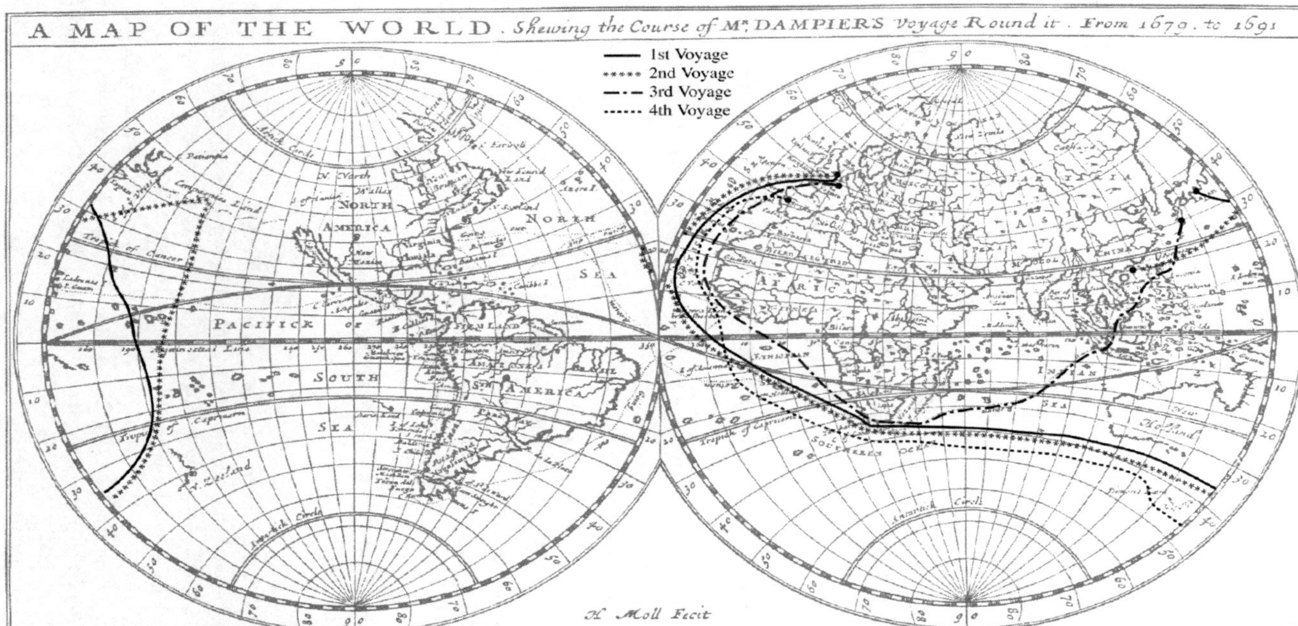

Fig. 1.3: Gulliver's homeward-bound voyages.

The Indian and Indonesian routes were the more usual ones to and from the Far East, but three times Gulliver is picked up by ships sailing back to Europe by a southern route, round the east and south coasts of the still largely unknown Australian continent, a route passing between Australia and Tasmania, the so-called New Holland and Van Diemen's Land of old maps (78, 148, 284) (fig. 1.3). According to Arthur E. Case, there was "general agreement" that this route was a possible one.[28] This view seems to be confirmed by older maps than Herman Moll's. A local one, *Hollandia Nova detecta 1644*, identifies at least two regions on the south coast mentioned on Gulliver's map of the Land of the Houyhnhnms: Land of Peter Nuyts and Land of Lewin.[29] On the other hand, it seems to have been a doubtful route for ships sailing west as, unlike Tasman, they would have had to sail against the trade winds and sea currents. The only identified navigator to have attempted this route is fictional Singleton, when Defoe has him sail round the still mysterious continent on his way to Ceylan (170–71): hardly a case of convincing evidence.[30] In Book IV, Gulliver is instructed by the merchants who employ him as captain to sail to Barbados and the Leeward Islands in the Caribbean and then, following in Dampier's footsteps, to "trade with the *Indians* in the *South-Sea*" (222) by way either of the Strait of Magellan or round Cape Horn, the route preferred by Dampier and those who followed his instructions. We are finally told that, apart from the four voyages narrated in his memoirs, Gulliver made a few others, in particular two to the Levant (19, 153), another frequent destination of merchantmen. All the outward voyages follow trade routes that one historian has called "the arteries of the [British] imperial body between 1650 and 1750."[31] On one of them, Gulliver even falls victim (154) to buccaneers like Captain Singleton, whose "piracy … was already a lucrative trade […] off the coasts of China and Japan."[32]

28 Arthur E. Case, *Four Essays on "Gulliver's Travels"*, 56.

29 The map is found in a travel anthology, *Relation de divers voyages curieux*, published in 1663 by the French scientist Melchisédech Thévenot. It is said to be a copy of one by Johan Blaeu from 1659.

30 See Arthur W. Secord, *Studies in the Narrative Method of Defoe*, 154–56.

31 Markus Rediker, *Between the Devil and the Deep Blue: Merchant Seamen, Pirates, and the Anglo-American Maritime World, 1700–1750* (Cambridge: Cambridge University Press, 1987), 21.

32 Douglas Chambers, *The Reinvention of the World: English Writing 1650–1750*, 24.

Swift's geography lays greater emphasis on overseas routes than on the European ones, yet the latter are not ignored. Within four days of his arrival in Amsterdam, Gulliver finds a passage back to London (217) and, in Lisbon, where Crusoe has a generous friend, he has no difficulty finding another passage once he has been convinced by his host that he should return to his country (288–89). This suggests regular sea links between European countries. Like *Captain Singleton* (164–66) and *A New Voyage round the World* (91–92), the *Travels* furthermore refers to the busy local trade plied between China, Japan, the Philippines and other offshore islands in the North Pacific. In Book III, Gulliver is entrusted with a sloop to trade between Tonkin and some neighbouring islands (154), as Dampier had once intended to do. The trading venture is cut short by Asian pirates as ruthless as Defoe's Captain Singleton, who rampaged in the same waters.[33] After leaving Laputa, Gulliver can take advantage of the frequent ships sailing between Balnibarbi and Luggnagg, and between Luggnagg and Japan (203, 215). These passages constitute so many plausible sea routes that contemporary readers may have added in their imagination to their mental maps of the world.

It is clear that, in its mapping of sea routes and emphasis on trade, Gulliver's geography is, like Defoe's, basically economic. In fact, it is again more economic than political, except, of course, in the pages in Book I recounting the difficult relations between Lilliput and Blefuscu, which feature the sea as a protective barrier. What is surprising in the *Travels* is that there is no mention of French ships, although they were active in the same waters. The English, the Dutch, and the Portuguese, three partners in the Great Alliance against France, seem to monopolize trade with the Far East, but the satire reminds us that, while the ships of the three nations were ready to help one another in emergencies at sea, they remained fierce rivals when it came to keeping their monopolies. Gulliver experiences it with the hostility of the Dutch in Japan, a hostility that nearly costs him his life or at least freedom (216–17). For Swift, trade is a competitive activity in which rivals give one another no quarter. He does not share Defoe's optimism about the beneficent influence of trade.

33 *A Supplement to the Voyage round the World. Describing the Countries of Tonquin, Achin, Malacca, etc., their Product, Inhabitants, Manners, Trade, Policy* (London: 1700), in *Dampier's Voyages*, ed. John Masefield (London: E. Grant Richards, 1906), 558, 591. The long delay of Gulliver's ship in Tonquin finds an explanation in Dampier, who reports a similar experience.

In his geographical discourse of sea routes, Swift proves more conscious than Defoe of what Anne Godlewska calls the simplified character of cartographic representations.[34] Swift's geographical discourse pushes their sketchy and diagrammatic character to extremes, like his maps. He can be said to oversimplify reality. He empties seas of their islands in order to make room for imagined ones or convey the impression of movements across vast expanses of water. A comparison between *A New Voyage* and "A Voyage to Brobdingnag" illustrates this. As seen above, in this novel, Defoe describes scrupulously, stage by stage, the progress of the ships around the world. Swift, on the other hand, with a bold and impatient stroke, has tempestuous winds send Gulliver's ship from Madagascar across the Indian Ocean to the east of the Moluccas (83), thus miraculously avoiding any contact with the barrier of islands formed by Sumatra, Java and the small islands separating the latter from Timor. He may have been thinking of the larger opening east of Java mentioned in *Captain Singleton*.[35] Then there is what one can only call the complex interaction between geographical discourse and maps. If, in Defoe's novels, maps were unnecessary and were only supplied as an afterthought, in Swift, maps and geographical discourse are woven in such a way as to create an alternative reality and produce a widely different picture of the world.

In the perspective of the present study, the fact that we do not know who had the idea of the maps in the *Travels* and had them engraved is not important. What must be remembered is that the maps were already there in the first edition and that they have since been faithfully reproduced in every new one.[36] For readers, they were, from the very first, part and parcel of Gulliver's geography and, as will appear below, they can be said to display the same teasing spirit as the narrative. On the one hand,

34 Anne Godlewska, "From Enlightenment Vision to Modern Science", 250.

35 *Captain Singleton*, 164: "As we now steer'd, our Course lay N. N. E. and we passed thus with a fair wind, thro' the Straight or Channel between the Island of Gilolo [Waiego?], and the Land of Nova Guinea, when we were soon in the open Sea or Ocean, on the South East of the Philippines, being the great *Pacifick*, or South Sea, where it may be said to join itself with the vast *Indian* Ocean."

36 The inclusion of the charts in the first edition may or may not have been one of the grounds for Swift's dissatisfaction with the printer and bookseller's work. What is sure is that Swift did not seize the opportunity of the Faulkner edition to drop the charts or have them revised, so that they have always been there. From the point of view of the reader, geography in the *Travels* is in the first place that of Gulliver, supposed author of his own portulan drafts, and not that of the satirist.

however fanciful they are, the maps were and still are used for their irresistible power of persuasion: they purport to be authoritative documents. Simultaneously, unlike Defoe's, they take advantage of the very imprecision of seventeenth- and eighteenth-century cartography and of the capacity of visual representations to create alternative worlds. One way or the other, the maps are clever instruments to manipulate the reader. Even if he is not the initiator of the maps, the satirist is able to play with them a teasing game of suggestions advanced and withdrawn.

Originally, their air of authority derived from several factors. One was that, in the course of the seventeenth century, seeing had become the predominant mode of apprehending the world.[37] As pictorial representations of the world, maps spread out the visible and observable and spatialized it. With the maps, the satire could capitalize on this new way of looking at reality. A second factor was that cartography synthesized the present state of knowledge of the world and was a collective undertaking, consequently credited with being trustworthy.[38] These factors, which are still valid today, have led a recent historian to conclude that in fiction "maps confer authority upon their makers or impose on their readers."[39] In the eighteenth century, they imposed on the readers all the more effectively. The maps then fit in with the general strategy of the satire consisting in convincing readers of the reality of things. It is the same strategy that in the text enlists actual people, cartographers like Nicolas Sanson d'Abbeville (1600–1667) (100) and Herman Moll (284), both authors of atlases and, since the Faulkner's edition of 1735, "Cousin Dampier" (5), the author of the genuine *New Voyage round the World* referred to in the introductory "Letter from Capt. Gulliver, to his Cousin Sympson", to claim factuality for Gulliver and his spurious discoveries.

37 David N. Livingstone, *The Geographical Tradition* (Oxford: Blackwell, 1992), 98: "the map subverted the traditional authority of the word in history and reportage in favour of what Alpers calls 'the new testimony of the eye.'" See also Svetlana Alpers, *The Art of Describing. Dutch Art in the Seventeenth Century* (Harmondsworth: Penguin, 1989), 159, and Charles Taylor, *Sources of Self. The Making of Modern Identity* (Cambridge: Harvard University Press, 1984), 144. All three emphasize the importance of visual representation.

38 To the extent that maps purport to convince readers of the actual existence of Gulliver and his islands, they belong to what, in *Jonathan Swift: The Fictions of the Satirist* (Bern: Peter Lang, 1991) I call "the framing fictions" of the satire.

39 Graham Huggan, *Territorial Dispute: Maps and Mapping Strategies in Contemporary Canadian and Australian Fiction* (Toronto: University of Toronto Press, 1994), 9.

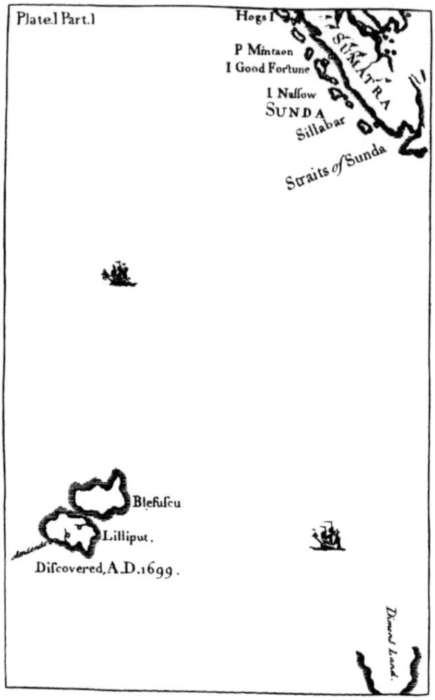

Fig. 1.4: Lilliput and Blefuscu.

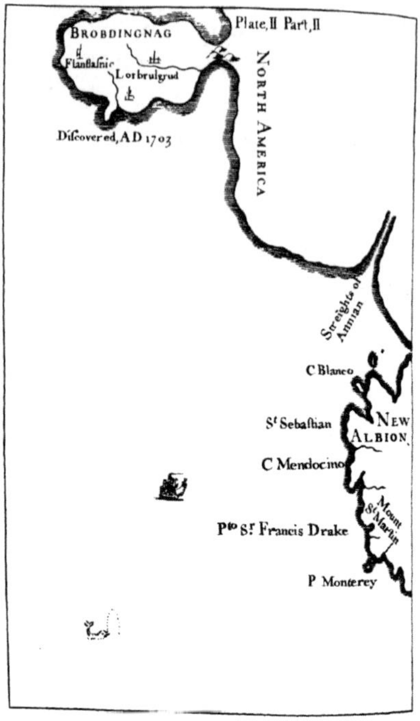

Fig. 1.5: Brobdingnag.

The maps in the *Travels* spare no details to underline the impression of accuracy and authority conveyed by the geographical discourse. There is always, in a corner of the maps or along one of their edges, the familiar outline of a continental coast or group of islands faithfully reproduced from existing maps, with the correct names of places. They show the mapmaker to be surprisingly well informed and quite up-to-date in his cartographic knowledge. The only problem for the modern reader is, even more than in Defoe, that many of these names of places are different from those found on modern maps. In the upper right hand-side corner of the map situating Lilliput and Blefuscu, the reader can see Sumatra (18) (fig. 1.4). The sea then is the Indian Ocean. The small islands of Mintaon, Good Fortune and Nassau off the coast of Sumatra and the Straits of Sunda between Sumatra and Java can be found on most maps of the time, beginning with the famous one by Johan Blaeu of 1663.[40]

40 *Le Grand Atlas; le monde au XVIIe siècle [compilé et édité par] Blaeu. Introduction et choix de cartes par John Cross; avant-propos de Peter Clarke, adaptation française de Irma Spinner* (Paris: Grund, 1992), 192–93.

Fig. 1.6: Laputa and Balnibarbi.

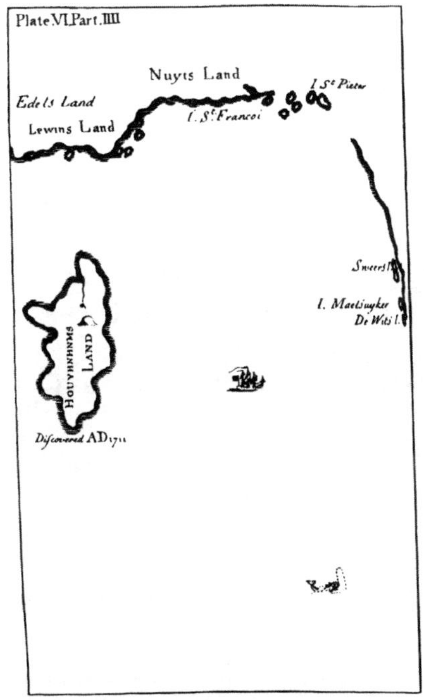

Fig. 1.7: Land of the Houyhnhnms.

The Isle of Hogs, which is not in Blaeu, becomes the Isle of Cochos on a map of 1753, but again l'Ile des Cochons on a French map of 1820.[41] The only inaccuracy, if it can be called one, concerns Sillabar. The name is usually engraved on Sumatra, not in the sea, as it does not refer to an island but to an English settlement. Dampier mentions it in his memoirs.[42]

The three other maps show the same attention paid to known geographical locations. On the map situating Brobdingnag (fig. 1.5), the names of places along the Californian coast, and several others, are listed in Henry Brigg's map of North America of 1625. The only exceptions are P. Monterey and the designation of California as New Albion.[43] Several

41 "Carte réduite de l'Océanie orientale depuis le Cap de Bonne Espérance jusqu'au Japon, de l'Académie de la marine, d'après Mamevilette, 1753", and "Carte de l'Asie de A.H. Brué, Paris, 1820".

42 William Dampier, *A New Voyage Round the World*, 272.

43 *Le Grand Atlas des explorations*, Encyclopaedia Universalis, 187.

of the same places are also mentioned in Blaeu, though not "Pt Sir Francis Drake": should one see here a nationalist bias? On the other hand, the depiction of the coast north of California is kept vague in all the maps, including Moll's map for Dampier's *New Voyage round the World*. Concerning the map prefixed to "A Voyage to Laputa" (fig. 1.6), the most detailed of all four, other existing maps attest to the existence of its place names. Iesso, C. Patience, Detroit of Uries, Sea of Corea, Yedo on Japan, Tonia, Bungo Island, Detroit of Diemen (Swift's Dimeris Strats), Tanaxima Island, Ongeluckick and Sondo Island (Swift's South Island), are all found on de l'Isle's map.[44] As for Plate 4, Part 4 (fig. 1.7), which situates Houyhnhnms Land, its unexpected accuracy has already been mentioned. Not only are "Edels Land", "Lewins Land and "Nuyts Land" the names of three areas along the Southern Coast of New Holland, namely Australia, that can also be found on seventeenth- and eighteenth-century maps but "I. St. Francot" seems to be a distortion of Isle of St. Franciset[45] and the Isle of St. Peter is also shown on these maps. As for those along what must be the coast of Van Diemen's Land, or Tasmania, the only one identified by the widespread maps is Madswyker Island. It is shown to lie further south, more exactly south-west, not due west, of Van Diemen's Land in the same "Carte de l'Océanie: cinquième partie du monde. Dessinée par Maire, Géographe". The names of places in this fourth map confirm that the coast of the Australian Great Bay must have been familiar to Dutch navigators earlier than has been assumed. It clearly suggests that the model for Gulliver's map must have been a chart by the cartographer Herman Moll, himself a native of Holland. For contemporaries, the inescapable conclusion could only have been that Gulliver's newly found lands existed, very much as did Crusoe's island.

Simultaneously, however, the satire cunningly undermines the reader's confidence in the reliability of the maps by frequent hints at the hypothetical nature of contemporary cartography. Swift's sceptical attitude to cartography is well known. As he put it in one of his poems, "On Poetry: A Rapsody":

44 "World Map of 1722 by Guillaume de L'Isle".

45 "World Map of 1722 by Guillaume de L'Isle" and "Carte de l'Océanie: cinquième partie du monde. Donnée par Maire, géographe, d'après Arrowsmith, Weyla et Faden. Paris, 1857".

So Geographers in Afric-maps
With Savage-Pictures fill their Gaps;
And o'er unhabitable Downs
Place Elephants for want of Towns.[46]

Deliberately or not, the utilization and nature of the maps in the *Travels* reflect a similar attitude. William Dampier in his memoirs and Defoe in the 1721 edition of *Robinson Crusoe* illustrate their narratives with recent and updated cartographers' maps. Out of the context of Moll's world map, of which they are fragments, Gulliver's maps, on the other hand, have the look of portulan charts which, as everyone knows, were tentative sketches of coastlines drawn by plattmakers for the benefit of seamen on the basis of voyagers' reports. The main maps in Dampier, that of the globe prefixed to the narrative and signed: "H. Moll Fecit" and those of Central America and the East Indies, as well as the map in the 1721 edition of *Robinson Crusoe*, are final 'scientific' products, with indications of latitudes and longitudes.[47] They vouch for the consistency of their respective geographies and indirectly confirm the truth of the report. By comparison, Gulliver's regional fragments of Moll's world present rough outlines for the newly found lands, a typical trait of some portulan charts. There is one exception, the map of the islands of Balnibarbi in Book III, which is modelled on Dampier's sketches of "The Islands of Pulo Condor" and "The Bashee Islands" and, paradoxically, looks more like the work of a navigator.[48] Be that as it may, the general impression is of approximate drafts.

The fact that the maps prefixed to each of the four books are presented separately and in a certain order and that they are not related to a map of the globe furthermore gives author and mapmaker ample scope to cheat with consistency. In Dampier and Defoe, the world map is a proof of the reliability of their respective geographies and hence of truth. In *Gulliver's Travels*, a portrait of the supposed author, in the style of the period, faces the title page instead of a world map: a subtle way of raising from the outset the question of the truthfulness of the

46 *Swift: Poetical Works*, ed. Herbert Davis (London: Oxford University Press, 1967), 574.

47 Christina Ljungberg, "Cartography and Fiction", *S: European Journal for Semiotic Studies* 15 (2003): 441.

48 William Dampier, *A New Voyage round the World*, 261.

supposed memoirs by drawing attention to an unidentifiable author. After that, the fact that the maps are scattered allows for changes of scales that conceal contradictory information. Thus the maps of Lilliput and Houyhnhnm Land show much the same area of the globe but, to make room for Lilliput and Blefuscu, the former fails to show the southern coast of New Holland or Australia north and north-west of Van Diemen's Land as the latter does (fig. 1.4 and fig. 1.7). Similarly, in order to make room for Brobdingnag, the map in Book II (fig. 1.5) fails to show Company's Land, which lies north-east of Japan on the map of "A Voyage to Laputa" (fig. 1.6). Had the maps been presented together or in a different order, the deception might have been more flagrant. Finally, not only does Brobdingnag replace the Company's Land of this map, but it is clearly drawn much smaller, although it should be bigger, further confirming the evidence of a change of scale.

As for the text itself, it underlines this uncomfortable ambiguity of habitually trustworthy artefacts presented as truthful untruths. When, in the text, the satirist makes fun of the lack of precision of maps in general, misgivings about the hero's geography are no longer out of place. In Book IV, he has Gulliver suggest that contemporary maps place New Holland, namely Australia, "at least three Degrees more to the *East* than it really is" (284). This "at least" is a humorous touch. Indeed, 3 degrees was a trifling error considering the state of cartography. The narrative informs the reader that Gulliver has communicated his views on the subject to Herman Moll himself, implying that Moll's maps should be corrected, as, actually, another error on his maps needed correcting, his engraving of Lower California as an island instead of a peninsula. The supposed inaccuracy concerning the exact location of Australia may then remind the attentive reader that to make room for Lilliput and Blefuscu, the first map had vaporized the still incompletely identified continent. What is certain is that the satirist not only makes full use of those areas of the globe that were still unexplored or little known in order to launch his hero into imaginary worlds, but his satire also raises the question of the speculative nature of contemporary cartography. In the case of Brobdingnag, Swift even turns the very fanciful hypotheses of cartographers back against them. There was a belief at the time that vast territories lay northeast of Japan, between Asia and North America. They had been given names Yeso and, as noted above, Company's Land or Land of Compania (Fig. 1.6). North of them, the map prefixed to "A Voyage to Laputa" bears the further precision, "Parts Unknown".[49] John Senex, Herman Moll's rival as a cartographer, was over the same years contributing to perpetuating the belief in the existence of this Company's Land.[50] Swift could also have got the idea of its existence from Sanson's older "Mappe-Monde" of 1641 (fig. 1.8). The coast between Oregon and Alaska being still unexplored, it was quite easy for a teasing satirist to take away some of the surface of this supposedly vast Company's land and of a largely unmapped North America so as to make them a suitable dwelling place for his giants. With the uncomfortable ambiguity of seemingly trustworthy representations presented as truthful untruths, the *Travels* poked fun at contemporary sailors and cartographers' claims to accuracy and suggested that maps were sometimes as fanciful as his own fictions. Seen this way, his inconsistent use of available geographical information from one voyage to another can be interpreted as one more knowing wink at the unreliability of Gulliver as narrator and at his boasted competence as a cartographer. There is nothing in this procedure to contradict what is generally known of Swift's use of multiple ironies. At any rate, the *Travels* shows an attitude to geographical information and cartographic representation diametrically opposed to Defoe's.

The iconic character of Gulliver's maps confirms this sceptical attitude to and ironic treatment of geographical material. Besides being teasing instruments of persuasion and deception about locations and the reliability of cartography, each map in the *Travels*, like all maps in fact, is "at once a synthesis of signs and a sign in itself: an instrument of depiction—of objects, events and places."[51] Introducing each voyage as they do, they claim to be helpful guides for the reader, while underlining the importance of the geographical background. "A Map of the World. Shewing the Course of Mr. Dampier's Voyage Round it. From 1679 to 1691" and "A Map of the World, on which is Delineated the Voyages of Robinson Cruso" announce in both cases that the stage of the narrator and central character's action is the real world.

49 A precision given in Moll's *New & Corrrect Map of the Whole World ... 1719.*

50 *Le Grand Atlas des explorations, Encyclopædia Universalis* (London: Harper, Collins, 1991), 187.

51 Denis Wood and John Fels, "Design on Signs: Myth and Meaning in Maps", *Cartographia* 23 (1985): 54.

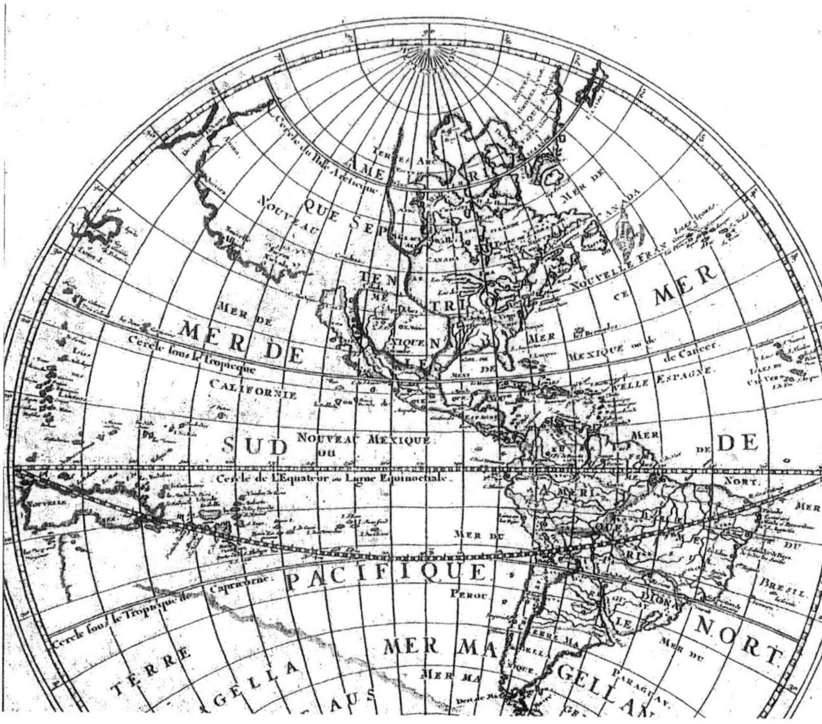

Fig. 1.8: On Sanson's map of the world, the Land of Campania or "Companies Land" that may have given Swift the idea of his Land of Brobdingnag is simply called "Nouveau", new. The landmass at the bottom of the South Seas is the hypothesised Terra Australis Incognita.

The dotted lines representing the course of the voyages lay the stress on the feat of the voyager: a circumnavigation of the globe in one case and extensive travels that come close to another voyage round the world in the other case.[52] The maps present the central figures as adventurers.[53] With Gulliver's four, or even five charts, the case is completely different. The repetitive pattern of which the charts form part is a reminder that Gulliver is a man who sails to earn his own living as surgeon or later captain in the merchant navy. The juxtaposition of the first or first two pages of the text and of the prefixed chart, on which the absence of any dotted line is conspicuous, further underlines the fact that Gulliver's ships never get to their intended destinations: far from it. The discrepancy between text and map also emphasises the hazards of navigation and trade. In some cases, the narrative leaves a gap in the geography. The tempest that sends Gulliver's ship to the inhospitable coasts of Brobdingnag disorientates the crew, who simply do not know where they are, and the giant eagle that allows Gulliver to escape drops his apartment box at a distance from the continent which it is impossible to compute. The most noteworthy case of discrepancy between text and map occurs in Book IV. When Gulliver leaves the land of the Houyhnhnms, he says that he heads east hoping to reach the southwest coast of New Holland and he eventually arrives at a coast stretching from south to north (284). This implies that the island of the Houyhnhnms lies west of the Australian continent. However, if one looks at the map sixty-three pages earlier, one finds their island to lie not east but south of the continent. Clearly, Gulliver does not know where the island of the Houyhnhnms is exactly and where he is sailing. He speaks of finding an uninhabited island where to spend the rest of his life and to reach Australia. In fact, he acts as irrationally as, in similar circumstances, Crusoe acts methodically and prudently. And the height of absurdity, it is this occasion that Gulliver chooses to propose amending Herman Moll's map of the area.

52 See the light dotted lines on the maps in Figs. 1 and 3.

53 The frontispiece of Defoe's *New Voyage* is allegorical. The shape of the South American continent within a hemisphere draws attention to the central role this part of the world plays in the second part of his narrative.

Second visual element: in Dampier and Defoe, the world maps show the continents as well as the oceans. They keep the correct kind of balance between the representation of land and sea. Because they are regional charts focusing on islands and coastal lines, the expanses of water dominate in Gulliver's four maps. The charts even suggest that these expanses of water extend well beyond the frame on two or three sides. His newly found lands thus look lost in unchartered waters. This is particularly evident in the map prefixed to Book I, where the coast of New Holland, north of Van Diemen's Land, is left out, making the Indian Ocean look the more boundless. It is no less striking in fig. 1.5, where the Brobdingnagian giants' peninsula is kept much smaller than it should be: in fact, its size is completely out of proportion with that described in the text.[54] The Pacific is immense, and it must have looked more formidable to Swift and his contemporaries than it does to us today. Dwarfing the land of the giants by comparison conveys this impression of a vast unchartered expanse of water. Maps invite the reader to trust his eyes.[55] On other maps, the lands and islands discovered are kept deliberately small, in other words relatively unimportant compared to known areas like Sumatra, the west coast of North America or even the southern coast of the still largely unknown Australian continent (fig.1.7). The vastness of the maritime spaces is further underlined by the then common cartographic device of adding pictures of ships and sea animals in the blank spaces. However big they may look to human eyes, the ships and whales are tiny on Gulliver's maps, much smaller than the ships and animals on the maps of older cartographers. The general impression is of vast seas where a few much more curious discoveries than that of Robinson Crusoe can still be made by accident.

At another level, the reading of the maps conveys a different and even contradictory message. Because they are regional, the maps compartmentalize space and produce a sense of order. They give the impression that space can be appropriated or at least made manageable by being mapped out. In this sense, they are representational models

54 John Robert Moore, "The Geography of *Gulliver's Travels*", 218.

55 On this problem of the size of the lands discovered by Gulliver, the visual aspect contradicts Arthur E. Case's reconstruction of the satire's geography according to modern realistic criteria in *Four Essays on "Gulliver's Travels"*, 54–55. He imagines a huge Brobdingnagian continent blocking the North Pacific.

concerned with domination, although not necessarily physical domination. They favour a detached attitude, including towards the sea, whose space obviously cannot be territorialized like land. One is reminded here of Johan Blaeu's comment in the dedication of his 12-volume *Grand Atlas* to Louis XIV: "maps," he writes, "enable us to contemplate at home and right before our eyes things that are farthest away."[56] There is no doubt that the reader adopting this detached attitude expects to be able to follow Gulliver's routes on the maps while reading the account of his adventures, as he can do with the light dotted lines on Dampier's and Defoe's maps of the world. Many have tried to show Gulliver's routes on the maps he provides, but not very successfully, because the satire is once more playing Swift's game of raising expectations only to disappoint them soon after. As already said, one can trace the routes to the destinations he never reaches but not those to and from the lands he visits. His satire is directed at the weak sides of contemporary representations of the world, at the frequent credulity of the public before them and at anyone tempted to believe that the sea as space can be dominated, be it only intellectually. His quizzical attitude contrasts with Defoe's enthusiastic embracing of the progress of geography. Swift's attitude is, in fact, the same that he manifests towards science in his portrayal of the Academy of Lagado in "A Voyage to Laputa".

Sailing around the world was a perilous job and the vision of the globe elaborated by the geographers could not make the public forget that the seas they mapped out were "first and foremost a space where the living takes place", as Henri Lefebvre puts it.[57] Least of all could fiction writers ignore the concrete, physical and moral experiences of sea-going people and, in particular, the experiences of the travellers whose memoirs had helped collect the information. At sea, Defoe's and Swift's characters are indeed confronted with a reality that bears little resemblance to the distancing of the geographer. For them, the sea is a space of collective undertakings and individual adventure. To put it simply, together individuals attempt, by means of navigation, to extend their control over

56 Amsterdam: 1663, 3. Quoted from David N. Livingstone, *The Geographical Tradition* (Oxford: Blackwell, 1992), 98.

57 "L'espace, c'est d'abord le lieu où se déroule le vécu," Henri Lefebvre, *La Production de l'espace*, 13.

its space. At the same time, individually, they try to make room for themselves among their fellow creatures by acquiring or keeping a social position and measure of self-sufficiency: their individual production of space, as the philosopher calls it. The above description of the two authors' cartographic representations of the sea and sea routes has highlighted the fact that their geographies are utilitarian and commercial, with a component of exploration. Beyond this very general resemblance, and that of their well-known stories of shipwrecks and discoveries of unknown islands, their depictions of their characters' experience of the sea are frequently so opposed that they look as though Swift had wanted to take the view contrary to Defoe's. Defoe is optimistic and minimizes the importance of dangers, rough seas, tempests, rocky coasts and buccaneering. After the two famous shipwrecks in *Robinson Crusoe*, in the first of which the hero is the only survivor (90–91) and, in the second, a dog (197), the author prefers to dwell on the control of sea routes. He usually has little to say about life at sea, as Tobias Smollett does later. Moll Flanders and Colonel Jacque cross the Atlantic several times safely, and there is nothing more to say about it. He focuses on what his heroes and heroines achieve by crossing the seas. For nearly all of the characters, after unpromising beginnings, sailing overseas is a step towards a better status. In *Robinson Crusoe*, the colonial link between Portugal and Brazil makes the hero a rich man without his doing anything about it. His plantation in South America prospers during the long years of his confinement on his uninhabited island and, when he arrives in Lisbon after his rescue, he can reap the profits from the sales of its produce, "Tobacco in Roll, and Sugar in Chests, besides Rum, Molossus, &c." (267). Moll Flanders, Colonel Jacque and the buccaneer, Captain Singleton, are not different. Although none of them is given any vocational training and all four knock about a great deal, they end up prospering overseas. It must be admitted that luck is a factor in their making good. The hero of *A New Voyage round the World* alone is a professional man: a sea captain who, with ideas and competence, turns his circumnavigation of the globe into a profitable collective undertaking. He is also Defoe's only fictional character who succeeds in extending man's control over sea routes to get access to world resources. But whether their success is individual or collective, all make room for themselves in this world and extend man's physical control over geographical space: they are literal producers of their own space.

Swift is as dubious of such success as Defoe is optimistic. He clearly takes pleasure in debunking such hopeful views as the novelist presents in his success stories. In the *Travels*, the sea as space of individual and collective activities tends to be one of suffering and death. At first sight, with four narrow escapes, Gulliver would seem, even more than Crusoe, to lead a charmed life. But this is only a superficial first impression. Unlike Defoe's heroes, he is a man dogged by bad luck to the very end. Not only do his ships never reach their intended destinations but, on each outward voyage, after being threatened with drowning or some other kind of death at sea, he is saved accidentally and, in order to leave the islands where he has found temporary refuge, he again risks his life at sea with no certainty that he will be able to get back to England. After his shipwreck in Book I, Gulliver manages to swim to the shores of Lilliput (21) and, on leaving Blefuscu at the end of his adventures among the Lilliputians, he entrusts his life to a probably leaky boat washed ashore after another shipwreck in order to flee the midgets who are plotting his death (77). Twice more, he finds himself sailing, not knowing exactly where, in a totally inadequate small craft: once when he is set adrift by pirates in a canoe, which allows him to reach Laputa (155), and a second time when he is forced to leave the Land of the Houyhnhnms in another primitive canoe, one made of stitched Yahoo skins (281). Twice he falls victim to pirates (155, 222), and twice his shipmates abandon him (85, 222). The mention of Madagascar in the second instance inevitably recalls Defoe's more lenient treatment of buccaneering in *Captain Singleton* and elsewhere. After being carried away from Brobdingnag by a giant eagle that drops him into the sea, Gulliver is eventually picked up by a homeward bound English ship (140–41, 143). On each voyage, the sea cuts short his professional activity and he returns to England as a mere passenger on another ship. Swift moreover insists that, even on the ships he sets out on in the first three voyages, Gulliver cannot for long carry on his professional activity as the ship's surgeon and, in the fourth one, as a sea captain, he fails to retain his authority, as opposed to the anonymous narrator in *A New Voyage*. In other words, he is never able to make room for himself in the sea trade. It is as though the sea each time rejected him and conveniently landed him physically on islands met on the way. Gulliver, the individual with professional skill, is not the detached geographer of his memoirs but a body tossed by the sea.

Yet, for all this, he is luckier than the shipmates with whom he sails from England. The death toll is heavy in the *Travels*, like that of the two shipwrecks in *Robinson Crusoe*. Through Swift's pattern of repetitions, the cruelty of the sea appears the greater. Defoe does not dwell on the cost in lives. Swift does. In Book I, Gulliver is the only survivor of the shipwreck. In Book II, his frightened companions, who leave him behind in Brobdingnag, apparently never return to England to report the existence of the giants and the loss of their ship's surgeon. At the beginning of Book IV, Swift adds the disappearance of still another ship, bound for America, like Gulliver's:

> On the 14th we met with Captain *Pocock* of *Bristol*, at *Tenariff* ... On the 16th he was parted from us by a Storm: I heard since my Return, that his Ship foundered, and none escaped, but one Cabbin-Boy." (221)

In the *Travels*, there are those who drown or disappear and there are those who fall ill and die aboard. During the long months at sea, sailors often go hungry (20), lack drinking water (85, 217), fall ill, have accidents and die as a result of the hard living conditions on board (217). The contrast with the regular re-supplying of the expedition in *A New Voyage* is striking. In Swift, the sea as scene of action or Marxist base or fundament for the production of space is untameable and hostile to man. And, as if this was not enough, human buccaneering makes the sea even more treacherous than it is by nature.

A scene of death and individual suffering, the sea is also a space of economic activities and there Swift's picture is equally dark. Regarding piracy again, Swift does not mitigate the crime by approaching the problem from the standpoint of chastened buccaneers like Captain Singleton. He is content to present the buccaneers' attempts to control sections of overseas routes as motivated by an egoistical desire for wealth or the fear of being prosecuted as criminals. The production of space consisting in developing overseas trade routes is no different: it is also motivated by egoism. Unlike Defoe, Swift does not dwell on the nature of the goods transported on European merchant ships.[58] There is no mention in the satire of pearls, gold, spices or calicoes. Swift prefers to

58 The only mention is that of logwood, "the central wood of an American tree (Hæmatoxylon Campechianum), used for making dye" (*Swift: Gulliver's Travels*, ed. Paul Turner, 362).

emphasize the fact that, thanks to overseas routes and trade, luxury goods flow into Europe to satisfy, at the cost of numberless lives, not essential needs but a taste for superfluous things. In this sense, the development of the narrative prepares the reader for the diatribe against luxury trade in Book IV, which includes the famous lines:

> I assured him [Gulliver's Houyhnhnm master], that this whole Globe of Earth must be at least three Times gone round, before one of our better Female *Yahoos* could get her Breakfast, or a Cup to put it in. (251–52)

The conjunction in the *Travels* of so much suffering and so many deaths with these images of a lady's breakfast and porcelain cups lends extra weight to the attack on the luxury trade. An informed contemporary reader could have spotted in these images, and in Book II, an indirect reference to Dampier's voyage to Tonkin, from where porcelain, silk and lacquer-work were imported[59] or, once more, to Defoe's description of trade in the Philippines and on the west coast of South America in *A New Voyage*.

What happens to Gulliver on each of his returns to Europe completes Swift's picture of the negative impact of the economic production of space. Unlike Crusoe, whose possessions overseas prosper in his absence,[60] Gulliver regularly comes back to Europe empty-handed and ends up as poor as when he started. He is never heard to get any salary from his employers on his return. How could he, considering the loss of his ships? He only once returns with something valuable: a red diamond, which he sells in England for "Eleven Hundred Pounds" (215). From his first voyage he brings back some Lilliputian cattle to exhibit "to many persons of quality and others" (79–80), but this is not something he and his family can live on. Back from Brobdingnag, the goods that he offers to leave behind "in Security for the Payment of [his] Freight" can only be a few worthless mementoes from his stay among the giants, items of little interest except to himself (148). His situation is still worse in the last two voyages. In Book III, he has to work and pay for his passage back to Europe: "understanding I was a Surgeon,

59 William Dampier, *A Supplement to the Voyage round the World*, 591.

60 Even after his escape from slavery Crusoe is not empty-handed. He manages to keep the boat in which he flees, as well as Xury, a young Moor who accompanies him, and various other goods. Selling them helps him settle as a planter in Brazil (81).

[the captain of the Dutch ship] was contented to take half the usual Rate, on Condition that I would serve him in the Way of my Calling." (217) Finally, in Book IV, he is even indebted to the Portuguese captain for the shirts on his back (288). It is clear for Swift that overseas trade may enrich a few merchants at home but definitely not the sailors, surgeons and officers who, all things considered, barely earn a living wage when they are lucky enough to survive the voyage: a far cry from Crusoe's or Captain Singleton's fates.

For Defoe, the sea is a link between countries. It opens natural routes to complete the exploration of the earth. The professional sea captain in *A New Voyage*, who is in this respect his most typical hero, pushes far south the exploration in the direction of the South Pole. He also dreams of finding a North-West and North-East Passage back to Europe and of completing the map of North America north of California:

> I had some Thoughts of steering away *North*, to try, what Land we might meet with to the *North East* of the Phillipines; and with Intent to have endeavour'd to make up into the Latitude of fifty or sixty Degrees, and have come about again to the *South*, between the Island of *California*, and the main land of *America*; in which Course, I did not question, meeting with extraordinary new Discoveries, and perhaps such as the Age might not expect to hear of, relating to the *Northern* World, and the Possibility of a Passage out of those Seas, either *East* or *West*, both which, I doubt not, would be found, if they were search'd after this Way; and which, for ought I know, remain undiscovered for want only of an Attempt being made by those Seas, where it would be easy to find, whether the *Tartarian* Seas are navigable or not? And whether *Nova Zembla* be an island, or joyn'd to the Main? Whether the Inlets of *Hudson's Bay* have any opening into the *West* Sea, and whether the vast Lakes from whence the great River of *Canada* is said to flow, have any Communication this Way, or not. (94–95)[61]

For the author himself and his fictitious narrator, routes furthermore give access to natural resources not always available in Europe. Defoe devotes a substantial portion of the *New Voyage* to the gold resources in the world and to possible ways for England to lay its hands on them as the Spaniards had done in South America. He also draws attention to what he considers empty lands suitable for agricultural development (212, 225–26).

61 It is important to remember here that Defoe's *New Voyage* reflects the geographical knowledge of the beginning of the century. The Tartarian Seas are the Arctic, and Nova Zembla is a land mass north of Japan, probably part of Siberia, already identified by explorers looking for the North-East Passage.

For Swift, the sea, with its dangers, is a barrier and it had better remain one. Where Defoe argues in favour of removing protectionist measures to promote open trading and in favour of colonizing empty lands, the satirist prefers to see the sea as a brake on economic exchanges and imperialistic designs, especially when they are unsound for their supposed beneficiaries, the European countries at the receiving end. Besides, the unpromising nature of Gulliver's new lands shows that he is sceptical about great developments in this direction. Perhaps he was thinking of the smallness of Crusoe's island. If there is one remark in Defoe to which the satire seems to take mocking exception, it is the allusion to the "extraordinary new Discoveries" that may still be made in the North Pacific in the above quotation. In the *Travels*, the sea functions as a protective barrier against the appetite of great nations and against the danger of certain cultures imposing their customs on others. The reader cannot help thinking of the Brobdingnagians and Laputans. Is it not fortunate, the satire asks, that the Laputans cannot with their flying island extend over other lands the domination they exert on Balnibarbi and that they cannot export their mad science and foolish way of life to other parts of the world (170–71)? And what if a king of Brobdingnag, whose kingdom is cut off from the rest of the world by rocky coasts and impassable mountains, became interested in conquering other countries, the satire seems to imply?

The sea protects small nations from big ones, the Lilliputians from Europeans, although not from their immediate neighbours in Blefuscu. Seas and oceans can prevent wars, act as a check upon undesirable trade and upon influences that might have a detrimental effect on remote populations. That such were Swift's views on the economic aspects of overseas trade and on colonization and its production of space receives confirmation from the attitude of the populations in the lands and islands discovered by his hero. The happier peoples are those that live on their own, exploit their own resources and ignore the rest of the world, the Brobdingnagians and the Houyhnhnms, but neither the Lilliputians nor the people in Japan, Laputa, Balnibarbi and Luggnagg, who have to take their neighbours into account. Swift's conclusion on the problem of economic exchanges comes close to those Voltaire expressed in *Candide*: let the countries cultivate their own gardens and leave the others alone. In *Gulliver's Travels*, one can sense Swift's scepticism before a Euro-centred globalisation of the world: an aspect of the satire that readers may be more sensitive to today than in the past. Defoe sees the world as the natural sphere of man's activities and oceans and seas as essential thoroughfares for economic development. In this, he expresses the mood of the time. Swift's view of the sea as an effective barrier demonstrates that he took exception to such views.

Repeatedly, in our examination of Defoe's and Swift's representations of the sea as geographical space and of the role it plays in their fictions, we have found the names of William Dampier, Woodes Rogers, Herman Moll and John Senex associated with theirs. References to their respective works have helped define Swift's and Defoe's visions of the sea and sea voyages as cartographic. Indeed, the two fiction writers are not simply indebted to Dampier for their stories of men stranded on unknown shores, as is commonly said.[62] Just as important is their debt to Dampier's discourse and cartographic vision and that of travel literature in general. There is also their debt to cartography itself. For the first time in English literature, with a global cartographic vision of seas, coasts, continents and islands, the scene of action of characters in works of fiction was the whole globe and its three oceans. With this cartographic vision and the added use of maps, *Gulliver's Travels* and *Robinson Crusoe* and, to a greater or lesser extent, Defoe's other novels about overseas adventures started a new trend in literature. This trend was conspicuous for works describing a world that is vast, open and exciting and portraying heroes always on the move, whose hazardous voyages were motivated by their curiosity to discover it. It is true that Gulliver is said to put out to sea driven by economic necessity, whereas Crusoe follows his adventurous disposition. But they are not so different. Gulliver speaks of his "insatiable Desire of seeing foreign Countries" (80), like Crusoe, who readily admits that he "had a mind to see the World" (68). Defoe's other main characters are no longer motivated by the same curiosity about the world, but his novels continue to make the most of the public's interest in remote countries and taste for geographical lore, as *Gulliver's Travels* does too. As for the object itself of this curiosity,

62 William Dampier, xxxii, 65–67. In his *Voyage*, Dampier reports the story of a Mosquito Indian. Later he was present at the rescue of Alexander Selkirk, whose long and lonely stay on one of the Juan Fernandez islands inspired the tale of Robinson Crusoe.

Gulliver's vision embraces nearly as much of the earth as Defoe's fiction and, with the extraordinary populations invented by the satirist, his *Travels* can be humorously said to surpass all the other contemporary genuine and fictitious travel accounts for its "extraordinary new Discoveries" and curiosities.

Swift's and Defoe's maritime geographies reflect and make the most of the prestige and new importance of overseas trade at the time to attract the reader. Their characters' travels are one and the same response to their economic environment. The main difference between them concerns their attitudes to trade and the exploitation of the world resources. For Defoe, in spite of Crusoe's many protestations to the contrary (60, 66, 68, 91), this vast world represents a chance and challenge that man should meet. Crusoe and the author's other heroes may at some point in their lives appear to be punished for reprehensible conduct or/and to be victims of society, but they fight their way to success. Defoe is confident that the world holds a better future in store for the individual who is not afraid to venture overseas. For Swift, seas are rather the gates to disillusions than to adventure to judge from Gulliver's four rescues, each as unprofitable as the previous one. Gulliver is forced by circumstances to embrace a career at sea and this career proves a failure. He is presented as a hapless victim, not a beneficiary, of the economic situation of the time. Furthermore, none of the places he discovers represents a new opening for English or European trade. As a result, it is not impossible to read *Gulliver's Travels* as a comment on Dampier, Woodes Rogers and Defoe. By intertextual implication, the satire brings out how unprofitable Dampier's circumnavigation of the globe had been for him in the first place, and for society, and how lucky he had been to survive at all. The satire is critical of the uses to which sea routes are put and, more generally, of the aims pursued by extending man's domination of the sea space. As for the contrast between Swift's and Dampier's works on the one hand, and *Robinson Crusoe* and Woodes Rogers on the other, it exposes Robinson's success story, and by analogy that of Defoe's other main characters as closer to wish-fulfilment than realism. Swift's economic geography in the *Travels* presents a point of view similar to that of his Irish tracts and pamphlets of the same decade and the next one when they defend Ireland against England's colonial policy.

The *Travels* has been simultaneously seen to be a satire about cartography. Defoe's use demonstrates his fascination with and complete assimilation of the geographical knowledge of his time. To Swift, this trust seems premature. The discordance between the cartographic viewpoint and the narrator's actual experience of the sea enables him to expose, besides the unreliable character of the narrator, the still important hypothetical and fanciful aspects of much contemporary cartography. To that extent, the divergence between the two writers is due to the state of geographical knowledge. However, Swift's criticism evidences a more radical mistrust. He concludes that the contemplation of maps is fascinating but its schematic oversimplifications are fallacious. They blind us as much as they enlighten us on the surface of the earth, its inhabitants and resources. The deception is particularly evident when one compares the virgin expanses of seas and oceans on maps, these blanks that small pictures of ships or whales attempt to enliven, with the hard reality of life at sea.

There is no evidence that Swift ever read Defoe's fiction. We only know that he did not hold the man in very high esteem. And yet, looking at the dates and similarities in their representations of the maritime space, at their geographies and at the flagrant oppositions of their views on trade and man's experience of the sea, it is difficult to escape the impression that Swift's satire was in some way a response to Defoe's views. Of course, it is not impossible that reading William Dampier, William Funnel, Woodes Rogers and Edward Cooke may have provoked the same reaction. Still, it is difficult to believe that the resemblances and disagreements between their works was purely accidental or due to their using the same geographical sources. The study of their land geography will confirm this impression. At this point, one can already note that the comparison between their representations of geographical space brings out little discussed fictional qualities in the two authors' works and enhances the interest of pages to which it is customary to pay little attention.

2. Gulliver's and Crusoe's Cartographic Representations of their Islands

With their maps, their textual mapping of overseas destinations of ships and their naming of places, known and unknown, *Gulliver's Travels*, *Robinson Crusoe* and Defoe's other fictions are conspicuous for their cartographic representations of the maritime space, trade and exploration. The same goes for their representations of land space. Their characters roam around islands or travel across continents much as they do across seas and oceans. In this case, however, the cartographic character of the representation is less evident. The reason for this is that the narrative discourse bears traces of the then rival art of geographical description referred to in the introduction, chorography, which was then losing ground to mapping but was well adapted to the depiction of local settings and their topography. In places, the vocabulary of chorography obscures the true cartographic nature of Swift's and Defoe's visions of space. Chorographic depictions, usually called prospects,[1] focused on the quality of the view offered and not on "the geometrical mode, or 'geometrica ratio'" of cartography (fig. 2.1).[2] They had behind them a solid tradition illustrated by profiles of towns, cities and estates. An English artist in this tradition was Wenceslas Hollar. A cartographer by trade, he also bore the suggestive title of "Scenographer and designer of prospects" (fig. 2.2).[3] In literature, Michael Drayton's descriptions of the counties of England in *The Poly-Olbion* (1598–1622), John Denham's detailing of a view over the Thames valley in "Cooper's Hill" (1642) and Pope's description of Windsor Forest in the eponymous poem (1713) are cited as instances of chorographic depictions of landscapes.

1 For some writers, prospects require an elevated viewpoint, which strict chorography does not.

2 Lucia Nuti, "The Mapped Views by Georg Hœfnagel: The Merchant's Eye, the Humanist's Eye", *Word and Image* 4 (1988): 545–46.

3 Douglas Chambers, *The Reinvention of the World: English Writing 1650–1750*, 25.

Fig. 2.1: Wenceslas Hollar, "Capture of Oppenheim (1688)". In this chorographic representation of the capture of the town, the point of view at ground level is that of a person standing behind the rump of the French general staff: a roguish wink of the scenographer?

Fig. 2.2: Wenceslas Hollar, "Edinburgh".

Such "topographical poems" as they are known, which combine accounts of "some particular landscape … with … historical retrospection or incidental meditation",[4] enjoyed an undeniable popularity until the nineteenth century. With their use of the dreamer's eye at ground level following the character's progress at a distance, John Bunyan's brief depictions of the hills that Christian climbs or of the crossroads on his way to the Celestial City in *The Pilgrim's Progress* could be said to be technically even more telling illustrations of chorographic representations than those presented in topographical poetry. There are even elements of chorography in genuine travel accounts written by navigators when they describe a coast or island descried from the ship. This is what a place on the Pacific coast that Woodes Rogers calls Tecames looked like from the sea when his ship approached it:

> The Land to the Northward, which is the Limits of the Bay of *Tecames*, is a long bluff high Point, and looks white down to the Water. The next Land to the Southward of *Tecames* is also white Cliffs, but not so high. I saw no Land on any part of the Shore, like those white Cliffs. Between them, which is about 3 Leagues, the Land is lower, full of Wood, and trimming inwards makes a small Bay, and the Village of *Tecames* lies in the Bottom, consisting of 7 Houses and a Church, all low built of split Bamboes, cover'd with Palmetto Leaves, and standing on Posts, with Hog-sties under them. These Houses have notch'd Pieces of Timber instead of Stairs to get up to'em. The Village lies close by the Water-Side, and may be seen when the Bay is open above 4 Leagues.[5]

As chorographic pictorial representations continued to be used, there was no reason why Defoe and Swift should not have relied on this mode of geographical representation. Did its attractive visual qualities not make it suitable for works of fiction intended to appeal to the imagination? The idea at least was not lost on later illustrators of their fictions. Yet, as will be seen below, Swift's and Defoe's descriptions resolutely put the stress on the more abstract and schematic mapping of lands and islands, on what C.S. Peirce, in his division of the class of icons defines as *diagrams* as opposed to *images*.[6] In the following pages, the analysis will mainly focus on the comparison of Crusoe's island

4 Samuel Johnson, *The Lives of the English Poets*, vol. I (London, 1799), 25.

5 *A Cruising Voyage round the World* (Santa Barbara: The Narrative Press, [1712], 2004), 139. The village is situated near the Ecuadorian Cape of San Francisco.

6 C. S. Peirce, *Philosophical Writings*, ed. J. Buchler (New York: Dover, 1955), 105.

and those that Gulliver discovers. The comparison between *Robinson Crusoe* and *Gulliver's Travels* highlights this cartographic character of their authors' treatment of space with clarity, and it will serve to complete the study, started in the first chapter, of the uncanny opposition of Defoe's and Swift's views on geography and of their characters' attitudes to space.

Swift is quite explicit about his indebtedness to cartography. Not only are there the maps, but his *Travels* also associates Gulliver with land surveys and mapping more often than with "prospects", a word that he nevertheless uses occasionally. In the third chapter of Book I, he writes that the Lilliputians intend to use Gulliver to get a better cartographic picture of their island: "an exact Survey of the Circumference of [their] Dominions" (44). In Book II, where the hero lives among giants and is least likely to get a bird's eye view of the country around him, the author first inserts an unexpected reference to the atlas of the French cartographer, Nicolas Sanson d'Abbeville (100), and then has his hero pace a gigantic "Royal Map" barefoot to study the layout of the country (112). Such allusions to the use and nature of cartography confirm the narrator's interest in what was becoming an established preference for the more scientific representation of the geography of the world.[7] They incidentally reinforce the impression that the maps in the *Travels* are Gulliver's. In his first novel, Defoe's life-long interest in cartography more simply finds expression in the detailed description of the island on which Crusoe finds refuge after the shipwreck. The description announces that of Moll Flanders' London, and reflects the cast of mind of its hero: Crusoe, like Gulliver, likes to measure space and is a born land surveyor.

It is at moments when the main characters are pictured looking around at their new surroundings that the two works of fiction may recall the chorographic model of geographical representation with its use of the viewer's vantage point. There is however a difference. They look around them with eyes influenced by the new geography. Gulliver's first view of the Lilliputian countryside, when he is allowed to stand up

7 Douglas Chambers dates the change from one dominant mode to the other back to the Restoration and the Great Fire of London in the second half of the previous century. See *The Reinvention of the World: English Writing 1650–1750*, 24.

and look around him, is a chorographic "prospect", and he does define it as a "painted Scene":

> When I found myself on my Feet, I looked about me, and must confess I never beheld a more entertaining Prospect. The Country round appeared like a continued Garden; and the inclosed Fields, which were generally Forty Foot square, resembled so many Beds of Flowers. These Fields were intermingled with Woods of half a Stang, and the tallest Trees, as I could judge, appeared to be seven Foot high. I viewed the Town on my left Hand, which looked like the painted Scene of a City in a Theatre. (29)

Yet, what is noticeable in this "prospect" is not so much its chorographic quality, a quality supposed to make the picture "spring out of [the page] in a tri-dimensional, lifelike appearance",[8] but the character's downward gaze at the scene below him, his measuring of his surroundings and the geometric nature of the space he looks at. All three flatten the topography. Gulliver's second general view of what goes on around him has an even more distinctive and amusing cartographic aspect. He sees the court of Lilliput "spread on the Ground" at his feet like a map:

> The Ladies and Courtiers were all most magnificently clad, so that the Spot they stood upon seemed to resemble a Petticoat spread on the Ground, embroidered with Figures of Gold and Silver. (31)

Crusoe's first view of his island from the top of a hill is more panoramic than Gulliver's but again aerial and more geometrized than chorographic. Like Defoe's other explorers in *Captain Singleton* and his own *New Voyage round the World*, he is mainly concerned with getting his own bearings in an unfamiliar geographical space:

> after I had with great Labour and Difficulty got to the Top [of the hill], I saw my Fate to my great Affliction, (*viz.*) that I was in an Island environ'd every Way with the Sea, no Land to be seen, except some Rocks which lay a great Way off, and two small Islands less than this, which lay about three Leagues to the West. (96)

The effect of this description is to make the island look lost in an immense sea, like Gulliver's islands on his maps. Crusoe adds that the hill on which he is standing belongs to a ridge spreading northwards (ibid).

8 Lucia Nuti, "The Mapped Views by Georg Hœfnagel: The Merchant's Eye, the Humanist's Eye", 545.

Later he records that from "the Brow of the Hill … [he] could see a great Way into the Valleys and Woods, which lay towards the *North-East* Part, and where the Island lay lowest …" (252). This information enables him to ascertain that he has landed somewhere near the eastern tip of the south coast.

Even where there are traces of landscape description, as when the voyager gets a first glimpse of a new place, the language is limited to listing basic coastline features, so that one hesitates to call such representations chorographic. This holds good for the fictions and genuine travel accounts. Woodes Rogers' description of Tecames Bay quoted above soon resumes a cartographic turn with heavy emphasis on measurements of distances and directions:

> There is a large Village about 4 Leagues off, where the Padre resides, and several *Indians* live between these Villages. The next River, 3 Leagues to the Northward, is very large, and call'd *Rio de las Esmeraldas*, but shoal'd; the Country about is thinly inhabited by *Indians*, Mulattoes and Samboes. By the Village of Tecames there is a River into which a boat may enter at half Tide; it flows here above 3 Fathom Water, the Flood runs to the Northward, and the Ebb to the Southward …[9]

In his description of the spot where Crusoe finds himself after miraculously surviving his shipwreck, Defoe has recourse to technical geographical vocabulary. Writing about his arrival on the island, he details "the Clifts [clefts] of the Shore" (90) that he manages to reach after the shipwreck, a "Neck or Inlet of Water… which was about half a Mile broad" two miles away westwards from where he landed (92), "the Mouth of a little River" with a "Creek", up which he pushes his raft and unloads the material he saves from the wreck (95, 96) and, finally, the "Hill" above the shore, which forms part of "a Ridge" stretching "northward" (96). In Brobdingnag, according to Gulliver's report, he and his shipmates land near "a small Neck of Land jutting out into the Sea, and a Creek too shallow to hold a Ship of above one hundred Tuns." (85) The coast appears to them merely "barren and rocky" (85). This dry geographical vocabulary, which offers little more than a rough picture of the layout of the place, does not allow the two descriptions of the heroes' arrival on an unknown island to become scenic.

9 *A Cruising Voyage round the World*, 139–40.

Since the first step in mapping out a newly discovered place is to situate it on the globe, both writers make frequent allusions to the points of the compass. Following the tradition of genuine travel accounts, they give as much precision as circumstances allow about the position of their heroes on the globe. Before the shipwreck, Crusoe's ship has been driven out of its course by adverse winds for more than twelve days. He remembers that, by the last computation made on board the day before the fatal event, the ship was still sailing in the Atlantic, somewhere between latitude 11° and 12° 18' north and longitude 22° west of Cape St. Augustino, the easternmost point of Brazil (87). The last tempest sends the ship scudding along westwards in the direction of the Caribbean Sea but Crusoe does not know how far. So, one of the first things he does after climbing the hill from which he gets his first view of his surroundings is again to determine his position. With the help of the date of his arrival close to the equinox, of the position of the sun at midday and one of the compasses or perhaps a quadrant saved from the wreck, he works out, as a good geographer, that the southeast tip of the island where he landed must lie at a latitude of 9° 22' north (104). On the other hand, he has no instrument and no way to measure the longitude, having lost count of the degrees during the tempest. He can at least surmise that he is now no longer in the middle of the Atlantic but closer to the Caribbean Islands, somewhere north of the South-American continent. This concern for the exact situation of his island never leaves him during his stay. Some time after his arrival, on a clear day, he can descry another larger island westward (137). From Friday's explanations and his own familiarity with world maps, he can eventually identify this other island west and northwest as Trinidad and deduce that his own island lies within sailing distance from the South American continent (216). Defoe thus suggests that Crusoe's island is one of the small islands in or near what contemporary maps like Herman Moll's "Map of the Middle Part of America" in Dampier's *New Voyage round the World* represent as the gulf of the Orinoco. Defoe was familiar with Moll's map and may have been inspired by it (fig. 2.3).[10]

10 William Dampier, *A New Voyage Round the World*, 26–27.

Fig. 2.3: Herman Moll's second map in Dampier's *New Voyage round the World* helps to situate Robinson's Island north of the mouth of the Orinoco and near Trinidad.

Fig. 2.4: The map of Crusoe's island in Defoe's *Serious Reflections* (1720).

In Book II of the *Travels*, when Gulliver's crew are approaching Brobdingnag after a similar tempest, they too conjecture that they are somewhere south of Great Tartary or the supposed Company's Land in the North Pacific according to contemporary maps,[11] so that they must steer due east to avoid it and the icefields. Defoe's and Swift's geographical positioning of the islands is, from the first, given a cartographic turn that implies that more systematic surveys will follow.

Equally cartographic are the two authors' representations of their islands. These are elaborated from the characters' movements and itineraries. This was the standard practice of cartographers as summed up by Henri Bouasse.[12] Inevitably, their descriptions differ in many ways. Crusoe's twenty-seven-year stay on the same small island leads to an elaborate treatment, characteristic of local geography, whereas Swift's depictions of Gulliver's shorter stays on ten different islands over several years is on the same scale of world geography as Dampier's or Edward Cooke's memoirs. Gulliver's maps largely make up for the scanty general descriptions in the text itself. In *Robinson Crusoe*, the narrator's description proceeds in such a manner that he can be said to construct a mental map of it for his own use, a map "representing the relations of parts in its designated referent".[13] Each new step adds a piece to the cartographic puzzle and the overall picture is quite different from that proposed by the frontispiece of the 1720 edition of *Serious Reflexions*, the only early attempt to draw a plan of the island (fig. 2.4). Unlike this awkward compromise between prospect and mapping, Defoe's geographical data focuses on the precise—precise as can be in Crusoe's circumstances— cartographic layout and topography of the island. As one would expect from the character, Crusoe's account of his movements abounds with indications of directions and distances. Although most of his time is spent meeting the pressing issue of ensuring his survival and improving the relative comfort of his dwellings, he never loses an opportunity to get better acquainted with his surroundings. He finds

11 Paul Turner (*Swift: Gulliver's Travels*, Oxford, Oxford University Press, 1971, 325, footnote 26) sees here a reference to Herman Moll's map of the world in his *Atlas Manuale* (1709), but Moll's earlier map of the globe in Dampier and other contemporary maps could have served the same purpose.

12 Henri Bouasse, *Géographie mathématique* (Paris: Delagrave, 1920), 287.

13 Stephen Bann, "The Truth in Mapping", *Word and Image* 4 (1988): 502.

time at intervals to explore the island systematically. He first ventures into the interior and then, to determine the contours and orientation of the island, he follows sections of the coastline on foot and even once launches a primitive craft to sail first up the east coast and then round its northern shore: a type of reconnoitring commonly practised by navigators and called coasting. Defoe's *New Voyage round the World* offers examples of this coasting to map out coastlines in its fictitious exploration of South Sea Islands.

His account first highlights the contrast between the two main coasts of the island. The eastern coast is the less hospitable one. As already pointed out, it stretches towards the northeast, is woody, cut by valleys (252) and exposed to tempests from the Atlantic and treacherous local currents. It is the coast where shipwrecks occur. Its hills are low and rocky. Near its southeast point, there is "a great Ledge of Rock" projecting into the sea, a dangerous spot for ships to negotiate, with treacherous currents and reefs, and a sandbank that just breaks the surface of the water (158–61, 168, 194). The western coast, on the other hand, is sheltered and pleasant, with hidden coves. The vegetation comes down to the water's edge (137–38) and most of the rich flora and fauna are found there. For the rest of the mapping, things are more complicated. Crusoe explains that, on the first expedition, he follows the western shore, walking "towards the *East* ... about twelve Miles" (139) to a point where he sets up a pole but without reaching the most northern point. On the second expedition, he has to sail first east from his main dwelling, past the dangerous reefs, and then northwards at a great distance from the east coast to reach "the northern Shore" (161). While trying to round it, he has to fight hard against another current and sail southwards and then westwards before he can get to the place where he had set up his pole (162). This walking eastwards to go to the northern end of the island and this sailing westwards coming from the northern point is confusing unless it is seen in the context of other precisions given about Crusoe's previous movements. One must remember that to reach the western shore from his main dwelling in the south, he has to walk "up the Creek" and keep on "due North" for a total of 8 miles or more to the spot where he has his second residence at the centre of the island (129–30) and, from there (137), he has to turn west to reach the beach. This suggests that the island is not only considerably longer than broad but that the west coast curves eastwards or becomes flatter on the said stretch of 12 mile. His third expedition leads him along the southern shore, three miles west of his main dwelling to the southwest end of the island, where many of the cannibals' feasts take place (178). As he says that the inlet where he first landed lies near the eastern end of this coast and that he must sail east from it to come to the dangerous ledge of reefs, this coast can hardly be more than five or six miles long. Concerning the northern shore, nothing is said about it and it must be short. Figures and references to the points of the compass suggest that the island stretches west south west and east north east and has an elongated form, something between egg-shaped and trapezoidal, with a particularly long western coast and short southern and northern shores. This picture ties in with what is reported of the sea currents, winds and tempests. The east coast is exposed to them, coming as they do from the west and southwest and from the middle and south Atlantic, whereas the west coast is sheltered and enjoys Caribbean calm (170).

The main trait of Crusoe's representation of the island topography is its division into specific areas. Here again, the reader is offered a map rather than descriptions of topographical landscapes. Crusoe even adds the plan of each of his dwelling places. Bruce McLeod has rightly noted that, in the course of his stay on the island, the hero divides it "into areas of labour, pleasure, knowledge, and defence."[14] In more geographical terms, these areas could be described as places of residence, exploitation and leisure activities. In the course of his exploration of the island, its woods and meadows in the valleys and hollows, Crusoe selects three places of residence. Near each of them he finds and can gather or grow precious food. Each of the three areas of residence and labour forms an entity. He builds his main dwelling, his "fort", as he calls it, below a rocky summit, on a projecting ledge "not above an Hundred Yards broad, and about twice as long" (101) and against the rock face— the surveyor's cast of mind manifests itself in the precision of the measurement. His tent faces north north west, protecting it from the Atlantic storms and rains. The ledge clearly overlooks the inlet up which he steers his rafts and unloads the goods saved from the wreck of

14 Bruce McLeod, *The Geography of Empire in English Literature* (Cambridge: Cambridge University Press, 1999), 198.

his ship. "Near [his] Dwelling … but a little farther within the Land, and upon lower Ground", he grows barley and rice (169). Going up the valley in which the river flows into the inlet below his main dwelling he discovers all sorts of useful plants. The spot at the centre of the island where he sets up his second residence, his "bower" and summer retreat, as he calls it, is situated in "an opening" sheltered from storms (130). It is also a watershed,

> where the Country seem[s] to descend to the West, and a little Spring of fresh Water which issue[s] out the Side of the Hill …, run[s] the other Way, that is due East … (131)

The place proves ideal for gathering fruit. He later has a second plantation of rice and barley near this bower (134) and builds an enclosure "adjoining" it for some of his tame goats (169). The third residence, discovered by accident northeast of his "bower", is at first sight an unpromising "little damp Piece of Ground in the Middle of [a] hollow and thick Woods" (177), but the grass is good and he erects there another enclosure for his tamed goats. It only becomes his third plantation (179) and third "Apartment" (187) when he discovers nearby a dry cave and realizes that the savages who occasionally visit the island are least likely to become aware of his presence when he is there. This third dwelling, where he eventually bakes his bread and stores ammunition, is somewhat too far north and out of the way to allow him to visit it as frequently as his other dwellings. On the other hand, its location, easily accessible from the east coast of the island, explains why, after the shipwreck of a Spanish ship off this coast, he decides to carry what he saves from the wreck to the cave instead of to his "fort". For his two leisure activities, Crusoe can sail on the quiet waters along the west coat, and he has the woods and unfarmed valleys for hunting. He constantly moves between the areas of residence, production and leisure activities, opening up paths between them. If the purpose of this geographical representation is eventually to situate the place where each important event takes place during his stay, so that the reader has no trouble following the characters' adventures, it must not be forgotten that his movements around the island are a central feature of his life, on a par with these important events, which moreover lead him to modify his circuits. Circuit is an appropriate term, a reminder that Crusoe is both a prisoner of the island dreaming of escape and a man who sees himself as its owner in charge of its supervision and defence against intruders.

Defoe's other novels present the same close ties between cartographic representation of space and action. In *Father Adventures*, *Captain Singleton*[15] and *A New Voyage*, when the characters go off in a spirit of discovery, they sail up and down rivers and walk across plains and mountain massifs, marking out their itineraries in relation to these topographical features like any good geographer. And they seldom travel without "all the necessary Instruments for Observation, … Perspective Glasses, Pocket Compasses, &" to mark out their routes and give an accurate geographical account of their movements.[16] Incidentally, in connection with this depiction of overland journeys across islands and continents, one is sometimes under the impression that the 90 degree angle of cartographic representation and the old way of symbolizing hills and mountains with scatterings of shaded molehills are at least partly responsible for Defoe's minimizing the importance of topographical obstacles in his desire to depict a world open to man. This comes out vividly when, in *A New Voyage*, he describes a party of explorers winding their ways between summits and precipices in the south Andes, passing effortlessly from one valley to another to reach the Argentinean Patagonia. Does not the narrator compare their impressions to those of people walking among the hills of southern England?

> We, who, as I said, had been for some Miles ascending upwards, were so high above them [the houses and gardens of an inhabited village] that the low Valley looked, as the Low Lands in *England* look below *Box-Hill* in *Surrey* …[17]

Defoe's comparison is a telling illustration of the often-mentioned fact that, at the beginning of the eighteenth century, geography had not yet acquired the necessary means to depict mountain landscapes, in particular great heights and depths.

To come back to Crusoe's island, it is more difficult to determine its size than its shape and layout. As noted above, the narrator here and

15 In a recent article, "Représentations d'Afrique dans le *Capitaine Singleton* de Daniel Defoe: une fiction cartographique" (*Dix-huitième siècle: l'Afrique* 44 (2012): 273–89) Erzsi Kukorelly also notes this close tie between cartographic representation of space and action.

16 *A New Voyage round the World*, 223.

17 *A New Voyage round the World*, 201.

there refers to the number of miles he walks between places, but the measuring of distances is far from being systematic. As already said, he has other things to do than measure the island. The reader has mostly to rely on the time it takes Crusoe to travel between places to estimate them and, here again, the evidence can be confusing. Crusoe more than once says that he uses his second residence as a halfway shelter in expeditions lasting several days (170). This would seem to suggest that it is at a considerable distance from his main dwelling. However, as he also keeps some tame goats there that he probably has to milk regularly, probably every day, the "bower" cannot be more than a couple of hours' walk from the "fort". At the end of his stay, when he describes how he disposes of the English mutineers he takes prisoners (256), he gives the reader to understand that, in an emergency, one can walk the distance between his "fort" and third "Apartment", the cave, in a matter of hours. Despite this lack of precision, there are sufficient indications to endow the island with, beside a shape, a sense of territory. Its depiction conveys the impression that Crusoe's island is, though not a kingdom as he claims (131), at least a princely estate. *Farther Adventures* confirms it, adding that it is large enough to support a small community of something like twenty tenants with their families, when Crusoe becomes their absentee landlord.[18]

The cartographic layout of the island presents it as an agricultural paradise with different areas favourable to different kinds of developments. Because much of the island is covered with trees and meadows, there is plenty of wood for cooking, building or making furniture and other utensils with. The wild goats that freely roam the woods and graze in the meadows provide Crusoe with meat, milk and hides, beside game to hunt. They can also be bred as domestic animals. The meadows along the river near Crusoe's main dwelling and the secluded spot among the hills in the north of the island are well suited to the purpose. Although the eastern coast is less hospitable and less luxuriant than the western

18 A central problem addressed by most studies is that of the hero's sense of ownership of the island, but it is seldom discussed in connection with Defoe's geography and handling of space. See Edith W. Clowes, "The Robinson Myth Reread in Postcolonial and Post-communist Modes", *Critique: Studies in Contemporary Fiction* 36 (1955): 145–59; Maximillian Novak, *Defoe and the Nature of Man* (London: Oxford University Press, 1963), 50–51; Carol Houlihan Flynn, *The Body in Swift and Defoe* (Cambridge: Cambridge University Press, 1990), 157.

one, its soil is as rich as everywhere else. Crusoe's cultures in front of his "fort" show it (142). In the centre of the island, where he sets up his summer bower, he finds

> the Country … so fresh, so green, so flourishing, every thing being in a constant Verdure, or Flourish of *Spring*, that it look[s] like a planted Garden. (131)

The lush vegetation produces a profusion of useful plants and fruit: wild tobacco, aloes, sugar cane, oranges and lemons, besides melons and grapes (130–31). Note that it is customary with Defoe to wax lyrical whenever he depicts soils favourable for farming or cattle breeding.[19] The only barren surfaces are the sandy beaches that the cannibals use for their human sacrifices but, even on them or in the woods, Crusoe can occasionally gather sea turtles or shoot a bird to vary his diet.

In his depiction of Crusoe's imagined Caribbean island, which shows the hero making the most of his fruitful surroundings, Defoe presents a geography of natural resources, a geography turned towards the opening up of new lands for farming: plantation is the word he likes to use. He can be said to put into fictional practice the recommendations made by voyagers like Woodes Rogers in their travel accounts. In this, *Robinson Crusoe* sets a pattern followed by his other novels dealing with voyages overseas. Each in turn enlarges on the subject. The exploitation of natural resources no longer concerns a small island: it extends to the whole world. Moll Flanders and Colonel Jacque prosper in the plantations of Maryland and Virginia. The same concern explains why the author's fiction comes back three times to the crossing of continents. In *Father Adventures*, Crusoe returns from the Far East by a continental trade route. In *Captain Singleton* and *A New Voyage*, Defoe specifies what Africa, the South Sea islands and Argentina have to offer: new types of animals, plants and fruit. Gold and silver are frequently mentioned, too. In his fiction, he does not only dream of open seas favouring a more efficient world trade but also of open continents with routes giving access to their natural resources. Obviously, letting natural resources run to waste jars with his protestant ethics. Perhaps a case of intended irony—or is it unconscious?—he has a Spaniard in the *New Voyage* lament the fact that his countrymen in South America do

19 See for example *A New Voyage round the World*, 129, 190 216, 258.

not bother to sift the gold washed down from the Andes[20] and that they leave the Costa Deserta, the Argentinean Patagonia unfarmed.[21] Ideas of overexploitation and exhaustion of natural resources never cross his mind, which may shock more today than it did then. The same *New Voyage* reports with approval bordering on complacency three cases of ruthless slaughter of animals by crews of ships: goats on one island, sea turtles and their eggs on another, and penguins on the Costa Deserta.[22] Defoe considers nature's bounty to be inexhaustible and generous to the enterprising man.

Highlighting as it does the plan of Crusoe's island and its immediate surroundings, Defoe's flat and geometrised picture is a typical illustration of the cartographic geography of the time and of its practical character. Practical it is in a double sense: it situates the events of the hero's life with respect to each other and it locates the resources he needs to survive. It serves the same purpose as portulan charts of newly discovered lands for navigators. It may be imprecise, even incomplete in several respects. The details of the cartographic puzzle scattered through the account of the hero's stay on the island present only a rough picture. Yet there is no doubt that, for Defoe's contemporaries, the cartographic picture, imperfect as it was, met the current standards of accuracy in the representation of remote places.

Swift's geography of Gulliver's islands and of the Brobdingnagian peninsula is more markedly two-dimensional than that of Crusoe's island. Largely owing to the world scale, it is stripped of unnecessary descriptive information whether pictorial or verbal. This has one literary drawback and one geographical advantage. At the same time as they propose simplified pictures of reality, maps and descriptions acquire some of the "scientific" authority that cartographic geometrizing confers on the representation of space.[23] The first object of the maps of Lilliput, Brobdingnag, Balnibarbi or the land of the Houyhnhnms has been seen

20 *A New Voyage round the World*, 173–75.

21 *A New Voyage round the World*, 210, 212.

22 *A New voyage round the World*, 142, 165, 226.

23 Lucia Nuti, "The Mapped Views by Georg Hœfnagel: The Merchant's Eye, the Humanist's Eye", 547: "when the mapper reaches a 90 degree angle (the zenith), the eye will loose its power again."

in the first chapter to situate them in the different seas of the globe and nothing has been omitted to persuade the reader of their existence. They are first of all credible eighteenth-century scale models. Their indebtedness to maritime cartography accounts for their focusing principally on the contours of coasts. It accounts for the fact that the diagrams of lands and islands either remain blanks or are highly stylised and sketchy, limited to the indication of a few mountains, rivers and cities. The symbols for topographical features are the usual lines for rivers and small shaded molehills for hills and mountains. The latter are found in representations of Sumatra, Brobdingnag, the land of the Houyhnhnms (figs. 1.4, 1.5, 1.6) and the supplementary map of Balnibarbi. When depicting mountain ranges, the small molehills are more or less geometrically distributed and show no awareness of the importance of a massif or of the heights of summits: another trait characteristic of old cartography. One map, the supplementary one of Balnibarbi in the middle of Book III (169), adds the stylised picture of trees, perhaps to illustrate the more fertile area of the island that Gulliver visits with his host, Lord Munodi, during his stay there (175–76).

One distinctive feature is that the maps in the *Travels* indicate the presence of towns. For these towns, the author of the maps had recourse to two different symbols: tiny dots or circles for the towns in Lilliput, Laputa and Balnibarbi (figs. 1.4, 1.5), and stylised town profiles in Brobdingnag and Luggnagg (figs. 1.5, 1.6). These town profiles were a survival of an earlier practice,[24] while the tiny circles were becoming the current signs as cartography moved away from pictorial quaintness towards greater geometrizing. In a few cases, the maps also bear the names of a place or two, like Sillabar in Plate I, Part I (fig. 1.4). Gulliver's maps do not show roads, but again this does not signify as cartography, and especially maritime cartography, had until recently ignored them.[25] This presence of towns confirms the impression that, to their respective populations, Gulliver's islands and his Brobdingnagian peninsula are of a size at least comparable to the British Isles or Japan: whole worlds in themselves.

24 See Juan Martinez's 1578 map of the Western Hemisphere, Henry Briggs's map of North America (1625) or Samuel Fritz's map of the Amazonian Basin (1707) in *Le Grand Atlas des explorations, Encyclopaedia Universalis* (London: Harper, Collins, 1991), 119, 118, 146.

25 See next chapter.

Gulliver's maps may be rudimentary because this was the practice in the drawing of charts, but the outlines of their coasts at least mark off their territories and endow them with distinctive shapes. These shapes in turn confer an identity on them, as Defoe's mapping of Crusoe's island lends it a form between that of an egg and trapezium. It must be remembered that, even though Gulliver's maps may indeed be indebted to Herman Moll's "New and Correct Map of the Whole World … 1719", his newly discovered lands on them are pure cartographic inventions. Thus, from the first look at the map prefixed to Book I, Lilliput and Blefuscu are identified by the reader as twin diagrams of much the same size and roundish shape, lying west of van Diemen's Land and separated from each other by a narrow channel—the strait that Gulliver later crosses to capture the Blefuscudian fleet (fig. 1.4). The Houyhnhnms' island is a larger diagram, the size of Tasmania, with something of an animal shape and two prominent capes on the north and northwest coast (fig. 1.7). The single hill in the north suggests that it is otherwise rather flat. Most importantly, it is the only island where there is no town. Brobdingnag is that bulbous excrescence of a spacious North American Continent lying far north in the Pacific Ocean and separated from the rest of the continent by a narrow mountainous isthmus. Luggnagg and Balnibarbi are again two inseparable diagrams, both of more compact shapes than Japan but more independent than Lilliput and Blefuscu. They are surrounded by many smaller islands, a detail that also helps identify them.

The maps are convincing enough in themselves and it is hardly surprising that, in spite of all the contradictions between them and the text they accompany, and in spite of the cartographic blunders highlighted in previous studies and referred to in the preceding chapter, readers should at first tend to trust their eyes, even though they may later raise an eyebrow when they come upon Gulliver's descriptions. At any rate, it is from the maps that they get their first impression of Lilliput, Brobdingnag, Laputa, Balnibarbi and the Houyhnhnms' island, and first impressions leave traces. One of these traces seems to be that readers usually accept that there is a double standard for the scale of things represented on the maps and in the text: in the first case, a scale that applies to the world picture that Swift shares with his contemporaries and us, and in the second a scale that applies once Gulliver is inside the discovered countries. On this delicate subject of the discrepancies and lack of consistency in the fiction, it is pointless trying to find rational explanations as Swift himself seems to have deliberately blurred both the time scale of the voyages and the spatial one by drawing red herrings like fantastical tempests, shipwrecks, abductions, or encounters with magicians whenever the hero crosses the line between the real world and the fairytale world of his imaginary lands.[26] The device belongs to the same strategy of equivocation referred to previously: it asserts and then throws doubt on the truthfulness of Gulliver's cartography. With this, the satirist once more turns to account what many in his days continued to view as the fanciful character of much cartography, more likely to feed daydreams than to provide information. After all, as seen above, Defoe's geography also lacks precision and displays inventiveness on occasion. Still, the mixture of realistic and fantastical geography is what makes the *Travels* so special compared with *Robinson Crusoe*. In Defoe's fiction, each place has, or at least could have, its proper location on actual maps of the world as it was known, even when, like Crusoe's island, or the Camerones River in *A New Voyage*, it does not exist. On the other hand, in Swift, Gulliver's islands only exist in so far as they appear on his maps.

Focusing too exclusively on the discrepancies has in effect obscured the complementary nature of text and image. To the case of Glubbdubdribb, in which the map can be said to correct the text, must be opposed the instances in which the maps shed light on the text. Some maps add a city here or a river there, and this is precious information. The map prefixed to "A Voyage to Laputa" (fig. 1.6) makes good narrative sense in situating Balnibarbi further northeast and Luggnagg closer to Japan, Luggnagg and Japan being, according to the regular sea traffic between them, in closer contact with each other than Balnibarbi with Luggnagg.[27] Besides, the distance between Balnibarbi and Luggnagg is greater than

26 In this respect, it could be argued that the title of Book III, "A Voyage to Laputa, Balnibarbi, Luggnagg, Glubbdubdrib, and Japan", and the map make better sense than the narrative (193, 203) when they situate Glubbdubdrib, the island of magicians, a place of obvious magic, closer to the return of Gulliver to Japan and the actual world.

27 At 46° north latitude, 183° longitude reckoned up east of London (155), Balnibarbi clearly lies northeast of Luggnagg with its 29° north latitude and 140° longitude (193). Luggnagg is also closer to Japan, which stretches 35°–40° north latitude and 160° longitude according to Moll's map.

between Luggnagg and Japan. The fact that the distance is greater fits in with the fact that Gulliver's passage between the first two lasts a month—it is true that his ship is caught by a storm (203)—whereas his passage between Luggnagg and Japan takes only fifteen days (215). Concerning Lilliput, the hillock behind which Gulliver hides on the northeast coast in order to observe the Blefuscudian fleet is not shown on the map, as probably too insignificant to be worth mentioning. On the other hand, there is the symbol of a town on a river, clearly a town with a harbour, where the king and his court reside when they come to watch Gulliver capture the enemy fleet across the strait. It makes sense. The king must have a residence on this coast. The harbour also accounts for the traffic between the twin countries. If a detail is missing it is the presence of a counterpart to this Lilliputian harbour, the equivalent symbol of a seaport on the opposite coast of Blefuscu.

While Crusoe divides his island into zones favourable to his different activities, Gulliver, like the explorers crossing continents in Defoe's other novels, either travels right through the islands he discovers or at least travels sufficiently round them to get an idea of the topography of the interior and of areas of economic importance. The narrative tells us that, in Book I, after his shipwreck, he reaches the southern shore of Lilliput. Because of the small size of the island, he then walks across it and over to Blefuscu, which he leaves from the northern shore after presumably another short walk. Lilliput and Blefuscu need not be much larger than Crusoe's island to feed a population of midgets twelve times smaller than their visitor. In Book III, Gulliver is picked up by the flying island at the western end of Balnibarbi and leaves from Maldonada at the eastern end, as shown on the supplementary map. The journey from Lagado, the capital, to the harbour seems to be an affair of several days, considering it necessitates "hiring two Mules and a Guide" (193). The scale on which the island and the neighbouring ones are drawn is that of Japan on the map. The size of Balnibarbi and Luggnagg is comparable to that of Japan. The same applies to that of the animal world of the Houyhnhnms off the coast of southern Australia in Book IV. In Brobdingnag, on the other hand, Gulliver follows the king in his "Progress to the South Coast of the Kingdom", the beginning of his majesty's complete tour. For the hero, this is another journey of consequence, undertaken by the king and queen's gigantic horses and carriages, capable of covering considerable distances in a day or two: a journey that Gulliver can only survive thanks to a "Travelling-Box", a specially built apartment for his comfort (139). This detail is enough to lend the peninsula of Brobdingnag a continental size quite different from that of other islands, a fact confirmed by the line of the coast of North America on the map. The text describes countries whose size is proportional to that of their inhabitants, each nation benefiting from an environment comparable to that of islands like Japan or the main British Isles.

At some point in the narrative of Gulliver's movements usually comes a general cartographic summing up of the information, like in Dampier or Woodes Rogers, and the account is on the whole what the reader expects after perusing the maps. The description of Brobdingnag illustrates this geographical descriptive procedure. It is composed in a style similar to that of Woodes Rogers's account of California quoted at the beginning of the chapter and gives the same type of geometrical and statistical, in other words cartographic, information.

> The Kingdom is a Peninsula, terminated to the North-east by a Ridge of Mountains thirty Miles high which are altogether impassable by Reason of the Volcanoes upon the Tops.[28] Neither do the most Learned know what sort of Mortals inhabit beyond those Mountains, or whether they be inhabited at all. On the three other Sides it is bounded by the Ocean. There is not one Sea-port in the whole Kingdom; and those Parts of the Coasts into which the Rivers issue, are so full of pointed Rocks, and the Sea generally so rough, that there is no venturing with the smallest of their Boats; so that these People [the Brobdingnagians] are wholly excluded from any Commerce with the rest of the World. But the large Rivers are full of Vessels, and abound with excellent Fish …
>
> The Country is well inhabited, for it contains fifty-one Cities, near an hundred walled Towns, and a great Number of Villages. To satisfy my curious Reader, it may be sufficient to describe *Lorbrulgrud*. This City stands upon almost two equal Parts on each Side the River that passes through. It contains above eighty thousand Houses. It is in Length three *Glonglungs* (which make about fifty four English Miles) and two and a half in Breadth … (111, 112)

The description of the country confirms the facts mentioned on the map, in particular with regard to the rivers, the cities and the mountain range that separates Brobdingnag from the rest of the North-American Continent. The inhospitable coast and impassable mountains—unlike

28 A reminiscence of the Andean volcanoes of voyagers' travel accounts found in *A New Voyage round the World*?

Defoe, Swift believes in impassable mountains—explain why the inhabitants of the country live isolated from the rest of the world. From the text, we learn that the name of the second city on the map, Flanflasnic, should be inscribed below that of Lorbrulgrud, above the symbol of the second town of the same size on the south coast, and not near the third symbol for a town on the north-west coast. As for the presence of a third city on the map, it highlights the importance of the urban population. The description of Laputa, the flying island, is similarly exhaustive. Other descriptions privilege the sampling of a characteristic landscape, as seen in the quotation of Gulliver's first impression of the countryside surrounding the Lilliputian capital at the beginning of this chapter.

In the *Travels*, virgin lands do not exist and, in its representation of land space, there is little information about available resources. Swift's geography is more demographic and less focused on natural resources than Defoe's. Gulliver's newly discovered lands and islands are all inhabited. All but one have cities, and these cities are densely populated. It would not be possible for a European to settle there and start a plantation, as Crusoe does in Brazil and Colonel Jacque in Virginia. Lilliput may be as small as Crusoe's island, but Gulliver immediately realizes, when he wakes up from his restoring sleep on the shore, that it swarms with tiny inhabitants. It is an anthill. In a matter of hours, its king calls up "Five Hundred Carpenters and Engineers" living near the spot where Gulliver landed to build a cart on which to carry him to the capital (26). The description of Brobdingnag quoted above speaks for itself: small as it looks on Gulliver's map, it can feed an equally numerous population and, what is more, a population of giants. Laputa, the flying island, is a crowded place too with its absent-minded philosophers, each of whom has a "flapper" in tow to rouse him from his speculations and a wife who needs a lover to console her from her husband's neglect. Only in the land of the Houyhnhnms is overcrowding avoided by a strict policy of eugenics and birth control. The demographic situation of these lands and islands has more in common with that of Europe, China or Japan, than with Defoe's Africa and Americas.

As for the way of life of their native populations, it exhibits various degrees of sophistication, of "civilisation": it is never primitive in the sense that Friday is primitive for Crusoe. The inhabitants of these countries live in rural societies at a stage of development not unlike that of contemporary Europe. Swift turns out to be highly conscious of the importance of the urban development of his time. In Lilliput, Brobdingnag, Laputa, Balnibarbi and Luggnagg, he depicts countries with large capitals and towns. Lilliput, Blefuscu, Balnibarbi and Luggnagg furthermore have seaports indicating the existence of commercial activities and exchanges. Clumegnig in Luggnagg and the unnamed town on the northern coast of Lilliput, opposite Blefuscu, are ports built upriver like Bordeaux, Bristol or London. All these urban centres are supplied by more or less efficient agricultural sectors. In Lilliput, Gulliver is told that the king has ordered the villages around the Metropolis to deliver meat, bread and wine for the sustenance of the giant, as they do for the inhabitants of the capital (32–33). Only an efficient agricultural system could meet such an extra burden. The description of Balnibarbi dwells on the contrast between the unproductive farming methods of the population and Lord Munodi's prosperous estate. On the Houyhnhnms' island, the layout of fields and meadows from the first evidences sound farming methods:

> The Land was divided by long Rows of Trees, not regularly planted, but naturally growing; there was great Plenty of Grass, and several Fields of Oats. (223)

In all the lands visited by Gulliver, the population exerts a pressure on the soil. Clearly, a growing population and the people moving to the urban centres had become an issue for Swift. The *Travels* comes back again and again to the subject of the capacity of local agriculture to supply the population with sufficient food. In this sense, the King of Brobdingnag's remark to the effect that "whoever could make two Ears of Corn, or two Blades of Grass to grow upon a Spot of Ground where only one grew before; would deserve better of Mankind, and do more essential Service to his Country, than the whole Race of Politicians put together" (135–36), is not the isolated thrust at the agricultural policy pursued in Ireland that it is often taken to be. The king's words are a central statement of the satire arising out of the overall geographical and demographic framework of the fiction. Swift is as concerned with agricultural production as is Defoe, but the idea that natural resources are unlimited and should be exploited and fetched from afar is foreign to him. He advocates their careful husbanding. Unlike the novelist, who is impatient to see remote resources put to use and comes close to condoning their ruthless exploitation, Swift seems to recommend improving local cultures, self-sufficiency and contentment with what one's country affords. This attitude is in keeping with his views on trade. In the 1720s, his perception of the phenomenon of urbanisation is forward-looking, as is his scepticism about the remedies that his contemporaries were trying to find to deal with this new social and economic situation: the opening of routes across continents, overseas trade, emigration, the development of English settlements on the American continent and agricultural experiments like those ridiculed in the episode of the Academy of Lagado. The often discussed resemblances between Gulliver's newly discovered countries and England are not just the source of satirical parallels: the *Travels* elaborates a genuine global geographical vision on the basis of its description of space. Its geography is not a mere fictional ploy of defamiliarization used to satirical ends. The mixture of fantasy and realism in its cartographic representation of the world may result in numerous discrepancies but, on the level of human geography, there is a striking consistency.

Defoe's and Swift's geographical visions of the world could hardly be more antithetical and it is no surprise that their heroes should undergo diametrically opposed experiences on their respective islands. The contrast is even starker than in the case of their experience of the sea: Crusoe conquers new spaces but Gulliver never gets a hold on the lands that he discovers. For Crusoe, Brazil and his island, one thinly inhabited, the other uninhabited, are open to European planters. In his European conception of things, he has a right to exploit the land as he thinks fit. Is he not, on his island, the only human to take advantage of its natural resources? No wonder that he soon comes to think of the place as his property. He reports that, on discovering the pleasant and fertile spot at the heart of the island, he was thrilled at the idea that "this valley was all [his] own". He adds:

> I was King and Lord of all this Country indefeasibly, and had a Right of Possession; and if I could convey it, I might have it in Inheritance, as compleatly as any Lord of a Mannor in *England*. (131)

At the beginning of his stay on the island, Crusoe's bird's eye view from the top of a hill had given him at most a general grasp of the place where he had landed. His mapping and dividing the territory into distinct areas progressively enables him to extend his physical control over its space and resources. Like Moll Flanders' and Colonel Jacque's America, the island offers him complete freedom of action, as long as he proves to be a clever and hard-working entrepreneur. After his rescue of Friday, of the latter's father and of a Spaniard from the cannibals, his sense of possession of the island becomes still more pronounced:

> My Island was now peopled, and I thought my self very rich in Subjects; and it was a merry Reflection which I frequently made, How like a King I look'd. First of all, the whole Country was my own meer Property; so that I had an undoubted Right of Dominion. *2dly*, My People were perfectly subjected: I was absolute Lord and Lawgiver; they all owed their Lives to me … (235)

The account of his twenty-seven years there concludes with his ownership of the island being recognized officially by a representative of a European power, namely the captain of the ship that brings him back to England. This conception of "l'espace vécu", of the living experience of space, which equates production of space with its use and possession, on the one hand reflects the contemporary notion that one's political and economic status depended on ownership of land[29] and, on the other, expresses the ideology of the period, which led to the English colonialist policy of land appropriation, a policy "predominantly concerned with securing rights not over peoples [like the Spanish] but over lands …, argu[ing] that the territories the settlers wished to appropriate were in some sense unoccupied."[30] In so far as Crusoe is represented as a hero who exploits untapped natural resources and considers that "all the good Things of this World, are no farther good to us, than they are for our Use" (152), his story could be considered an early classic of American literature in that it develops the theme of the frontier. He is a man who, like the author—now the words are Defoe's—

> cannot believe that God ever design'd the Riches *of* the World to be useless *to* the World; that the Gold, the Silver, the Diamonds, and other Species of such Immense Worth and Value, was ever created in the Bowels of the Mountains, and the most

29 Wolfram Schmidgen, *Eighteenth-Century Fiction and the Law of Property* (Cambridge: Cambridge University Press, 2000), 73–76.

30 Anthony Padgen, "The Struggle for Legitimacy and the Image of Empire in the Atlantic to c. 1700", *The Origins of Empire*, vol. I, ed. Nicholas Canny, *The Oxford History of Empire* (London: Oxford University Press, 1988), 41.

hidden Parts of the World to lye buried there, and remain unprofitable, till they come to be melted down again in the *General Conflagration* …

And the same applies to land abounding in "Corn and Cattle".[31] One can easily imagine the author of *Candide*'s grin if ever his eyes fell on this declaration of faith.

Curiously enough, the cartographic puzzle of the island and its division into areas used for specific purposes is completed about the time when, in the narrative, the first sign of a threat from the outside world, the ominous human footprint, appears. The danger heightens Crusoe's sense of his ownership. Bruce McLeod shrewdly remarks that the character's "compartmentalization" of the island, which amounts to his appropriation of it, is what "giv[es] form not just to himself but also to the Other, when or if they arrive."[32] What this remark implies is that Crusoe's sense of identity is at least as deeply connected with his ownership of the island as with his horror of cannibalism.[33] He can be considered a perfect embodiment of the Marxist producer and organizer of geographical space, and his story can be read as an illustration of what Swift's satire exposes as a shift in meaning, and confusion, between mapping and possessing. It never occurs to Crusoe that the natives have a right to use the island as they please and that they may resent his appropriating the place. In his production of space, Crusoe, like Moll Flanders, Colonel Jacque or any British landowner overseas, thinks in terms of personal property rather than in terms of colonial settlement. For Crusoe, the situation is simple: no one lives on the island permanently. The cannibals only come sporadically to sacrifice human victims on its barren sandy shores. For him, the island is unoccupied so that he feels entitled to exploit the fertile space. Besides, has he not "mixed his *Labour* with [it]; and joyned to it something that is his own, and thereby made it his Property", as John Locke puts it in his *Second Treatise of Government,*

31 Daniel Defoe, *A General History of Discoveries and Improvements (1725–27)*, 26, 27. The "General Conflagration" is a reference to the end of the world.

32 Bruce McLeod, *The Geography of Empire in English Literature*, 198.

33 Wolfram Schmigden, *Eighteenth-Century Fiction and the Law of Property*, 32–62. Like Schmigden, Brett C. McInelly describes the psychological phenomenon as one of "expanding Self" in "Expanding Empires, Expanding Selves: Colonialism, the Novel, and *Robinson Crusoe*", *SNNTS*, 35 (2003): 1–21.

1689–90?[34] He even goes further: he ends up expelling the natives from the island unless, like Friday, they accept to change their way of life and serve his purpose. This attitude has a long history and one that is not exclusively English. As a journalist has recently written: "primitive is what you call indigenous people … when you want to grab their land."[35] Defoe shows little concern for the natives, whom he looks down on. In *Robinson Crusoe*, he describes with exemplary clarity a situation that was recurrent in the history of colonialism and which was a source of mutual lack of understanding between Europeans and native populations. His handling of geography in this first novel highlights the problem more crudely than *Moll Flanders* and *Colonel Jacque* do.

It is possible to go further and argue that Crusoe first takes control of the island and then "peoples" it to further his ends. In his eyes, human relationships are subordinated to the ownership of land. Friday, whom he saves from death and befriends, becomes in effect his slave, as Xury had been after their leaving Morocco, although the boy had helped him escape slavery. The three British mutineers and sixteen Spaniards whom he leaves in charge of the island must see to its development. They are settled there to continue its exploitation. In *The Farther Adventures*, Crusoe refers to them as his "tenants, but tenants at will".[36] To make sure the island continues in his possession, he later even sends "seven Women, being such as [he] found proper for Service, or for Wives to such as would take them." (284) The way all this is expressed speaks volumes. Defoe's character may be a thoughtful owner, who cares about the welfare of the people working for him, but he manages them like his land.[37] Maximillian Novak says more bluntly:

34 John Locke, *Two Treatises of Government: A Critical Edition,* ed. Peter Laslett (Cambridge: Cambridge University Press, [1960] 1988), 288.

35 George Monbiot, "Who belongs to the past age", *Guardian Weekly*, March 31–April 6 2006, 16. The article was inspired by the eviction of the Kalahari bushmen from their ancestral lands and their resettlement in camps by order of the Botswana government. History continues to repeat itself.

36 *The Farther Adventures of Robinson Crusoe (1719)*, *The Novels of Daniel Defoe*, ed. W.R. Owens (London: Pickering and Chato, 2008), 205.

37 In *The Farther Adventures*, the rough plan of the different areas of production is doubled by a plan of the distribution of the people that Crusoe leaves behind on the island (Maximillian E. Novak, *Defoe and the Nature of Man*, 50–56).

Crusoe "tends to treat people like objects".[38] His handling of space and people in his fiction sheds dark light on the ethics of social relations and relations between peoples.

Compared to the genuine travel accounts of his contemporaries, Defoe exaggerates the availability of vacant lands to European settlers. Native populations can be discounted altogether when it comes to appropriating their lands. Swift must have sensed that there was something wrong about such an attitude and policy, and in his satire he goes to the other extreme. It never occurs to Gulliver that the lands in which it is his fate to be cast ashore could be uninhabited. To Defoe's empty American continent, the *Travels* opposes an imagined Far East of densely populated countries and old cultures, like China, and like the Japan of his own *Account of the Court and Empire of Japan* composed in the manner of Montesquieu's *Lettres Persanes*. He invents lands with civilisations comparable to that of Europe, countries whose populations could only resent the European attitude of grabbing their land: countries where what takes place in *Robinson Crusoe* cannot happen. The satirist anticipates Marx himself in his awareness that people produce space "but not just as they please or in circumstances of their own choosing."[39] His cartographic representation of the world can be credited with questioning the existence of uninhabited virgin lands—still an issue today as can be seen with what is happening to the Amazonian forest, that of Borneo, or in Botswana. He questions the legitimacy of laying claims on land that belongs to native populations. In this respect, Swift's human geography anticipates Diderot's *Supplément au voyage de Bougainville* (1772) and supports the views of those who contend that *Gulliver's Travels* includes an attack on colonialism. In the *Travels*, there can be no confusion between mapping and possessing. The

38 Maximillian E. Novak, *Daniel Defoe, Master of Fictions* (London: Oxford University Press, 2001), 245; see also 248, and Ian Watt, *The Rise of the Novel* (London: Chatto and Windus, 1957, 70). Another way of assessing Crusoe's attitude would be to say that, as self-styled monarch of his island, he oddly displays the same absolutist way of treating his 'subjects' that had made the Stuarts unpopular.

39 This is how Derek Gregory in *Geographical Imaginations* (Oxford: Blackwell, 1994, 110) sums up Karl Marx's view of history expressed in *Eighteenth Brumaire*. The quotation, which concerns history, is adapted here to geography on the assumption that for Marx and Marxists the two fields are intimately connected.

inhabitants, landowners and farmers of Gulliver's newly discovered lands are in full possession of their territories. Abandoned by his shipmates on the rocky coast of Brobdingnag, Gulliver climbs to the top of a knoll that gives him a "prospect" of the countryside. He finds it "fully cultivated" (85). After being set ashore on the Houyhnhnms' island, he walks away from the shore in search of some human presence:

> I walked very circumspectly for fear of being surprised, or suddenly shot with an Arrow from behind, or on either Side. I fell into a beaten Road, where I saw many Tracks of human Feet, and some of Cows, but most of Horses. (223)

This time, the landowners are horses.

The human geography and the narrator's account of his living experience of space in a foreign country combine to make the encounters with their populations the focus of our attention. Each of these populations confronts Gulliver with different versions of the Other. It is all too customary in criticism of Swift's work to discuss the Other as an embodiment of the self. This is regrettable, because it amounts to ignoring a central aspect of the fiction of the travel account: its geography precisely. Before becoming this embodiment of the self, the Other is always a foreigner, a rational being who looks, behaves, dresses differently and holds other values. Because this foreigner is in control of all the space and appropriation of land is out of the question, Gulliver must develop other ways of producing his space: he must find a place, his place among his hosts: a place these will be ready to grant him. The task before him is complex and the way less smooth than Crusoe's on his island, or Moll's or Jacque's in Maryland and Virginia.

In each of the four "Voyages", Gulliver reaches an unknown shore destitute, and three times with only a few clothes on his back. In Lilliput and Laputa, he arrives physically exhausted and half starved. In Brobdingnag and the land of the Houyhnhnms, his shipmates leave him behind, a Crusoe without a wreck to supply him with tools and weapons. He is fully dependent on the native populations for his bare survival. The theme of reliance on the natives' help is common in genuine travel accounts but Swift gives it an extreme turn. His satire dwells as insistently on the newcomer's dependence on others as Defoe's novels do on their main characters' self-reliance and aspiration to autonomy. Of prime importance in the *Travels* are the acquaintances that Gulliver can make. He must first establish elementary contacts: ask for food, drink and protection. This social and economic dependence of the individual on others, who are the foreign Other, is a basic trait of Swift's conception of elementary social relationships with the people of remote nations. For a longer stay, what Gulliver needs and looks for are relations of a higher order with his hosts: relations based on their accepting him and his gaining their respect. But this is no easy task in a world where repeatedly one comes among people completely different from oneself. Furthermore, at this early stage of acquaintance, the individual's production of space in the form of relationships of a psychologically more rewarding nature can only develop within his hosts' existing social network, and the higher up in the social scale, the more useful —or the more treacherous—these relationships can prove to be: a typical satirical thrust of the author.

Whether in Lilliput, Brobdingnag, Laputa or in the land of the Houyhnhnms, Gulliver tries hard and humbly to convince his hosts that he is like them, however different he looks. Unlike Crusoe, who makes Friday learn English to communicate with him, he learns their languages. He is ever ready to make himself useful: a recurrent source of comic of situations in the satire. Among the Houyhnhnms, he adopts the horses' values, their virtuous way of life, and goes so far as to imitate their way of speaking and walking. For a while, he can believe that he will be able to carve a niche for himself among his hosts but soon realizes that he is not welcome: that assimilation is impossible. Each exploration ends in the same way. The Lilliputians are at first impressed by his stature and think that they can make use of him as their cartographer and an ally against their enemies from Blefuscu. But they soon tire of his cumbersome presence and want to get rid of him. To save his life, he must put to sea again. In Brobdingnag, he tries in vain to convince the giants, and especially the king and his court, that he is not just a doll or an amusing mechanical toy, but they do not take him seriously for what he is, a rational human being like them. He is afraid that he and his descendants, if he should have any, might end their life in cages "like tame Canary Birds" (139). His accidental escape comes to him as a relief. Throughout the third voyage, Gulliver is confronted with queer sorts of humans. Laputans are so lost in their speculations that they show only fleeting interest in his presence and then ignore him. The academicians of Lagado's experiments and the Balnibarbians' unproductive economy intrigue him up to a point, although he dislikes their way of life. The Struldbruggs of Luggnagg and the ghosts of Glubbdubdribb only briefly fascinate him. At the end of his visit, he is glad to get away from them all, to reach Japan and find a Dutch ship about to sail back to Europe, even though he takes great risks in doing so, as no Europeans but the Dutch are allowed to set foot in Japan. On the Houyhnhnms' island, the Yahoos, who are humans, prove hostile. Only among the rational horses does he feel happy, but they will not let him stay on their island. He is banished. At the very end of the *Travels*, as though to hammer the point, Swift adds one more instance of rejection. After leaving the Houyhnhnms' Island, Gulliver reaches another island where he hopes to spend the rest of his life in lonely meditation but naked savages drive him off (284).[40] Each of Gulliver's adventures ends up with what can only be considered his exclusion. It is as a destitute body washed ashore or left stranded that he arrives in a new country, and it is as a rejected human being of European stock that he is thrown back into the sea in what is each time a narrow escape. If there is anything post-colonial in the *Travels*, it is the way the Other looks upon the main character. Swift's human geography concludes on the same sceptical note as his economic geography: like the Japanese in Book III, native populations around the world surely much prefer being left alone. People from different backgrounds do not mix easily. Physical and cultural differences may well be superficial, but they are a factor of exclusion.

Crusoe's production of space in terms of control over land and people fills him with a sense of his own importance. The same is true of Defoe's other main characters. Gulliver's confrontation with a bewildering range of natives different from him confuses him. This diversity is not simply one of outward appearances. More mystifying is the variety of natures and cultures. The small Lilliputians are clever, practical, active and pretty but treacherous; the Brobdingnagians have a somewhat coarse physique and sensibility, even though their king is a wise man. The geometrically cut clothes the Laputans wear are intended to make them look like pure intellects, who live mostly disconnected from reality. The academicians of Lagado strike one as unkempt but

40 Defoe imagines a similar episode in the South Sea, where the men of his expedition round the world are welcomed with bows and arrows (*A New Voyage round the World*, 133).

inventive. Most surprising of all are the Yahoos and the Houyhnhnms. The Yahoos share the "human form divine" but are filthy, lustful and behave like hordes of nasty monkeys, and the Houyhnhnms, their masters, who *are* animals, are comely, kind, sensible and behave rationally. Sometimes the Other is superior to Gulliver physically or intellectually and sometimes inferior. The Other may be endowed with a variety of qualities or infirmities. All this makes the visitor's attempts to render himself acceptable to his hosts and impose the image he would like them to have of him difficult. Their negative reactions can only have a debilitating effect on his sense of his own worth. But what is ultimately more confusing for Gulliver than this bewildering human diversity, what is downright mortifying is that the Other eventually considers his presence undesirable. Swift sounds sceptical about the readiness of the Other to become "mon semblable, mon frère" and about the possibility of fruitful intercultural relations and mutual acceptance between peoples with different histories, traditions and levels of development. He goes so far as to suggest that these relations can be traumatic. Faced with such an assortment of human and rational animals, who reject, despise or ignore him, his Gulliver cannot, to use Bruce McLeod's formula, give form to himself, or his self, whether by a clear sense of his difference or likeness, and he ends up mentally unbalanced. Far from giving him a sense of his own worth, Gulliver's contacts with other nations turn into a personal nightmare. Such is the conclusion of Swift's geographic fable.

As fictions about people and peoples, *Robinson Crusoe* and *Gulliver's Travels* touch upon the subject of human geography and international relations. In their interpretations, the two writers draw upon opposed ideologies, although both works reflect views and attitudes expressed in genuine travel accounts. The views of Defoe/ Crusoe reflect the transition from the ideology of the traditional land-owner to that of the new capitalist, who subordinates the worker to the requirements of production. Defoe's division of mankind into productive civilized Europeans and non-productive uncivilized non-Europeans expresses a commonplace view among his contemporaries. To this binary division, the satirist opposes an unexpected multiple division into what could be termed species: giants, midgets, average-sized humans, common horses and rational horses. Why species? Perhaps simply because species, as the twenty-first century has come to realize, are more neutral categories than races, as can be verified with the numerous breeds of dogs. As for societies, Swift offers a more balanced view of the situation in the world than Defoe in *Robinson Crusoe*: he describes a wide range of them, from the most primitive to the most sophisticated.[41] All present some qualities, or strong points, and a great many weaknesses and shortcomings. There is only one omission. The *Travels* makes no mention of cannibals. Swift saved them for *A Modest Proposal*, his satire on England and Ireland. *Gulliver's Travels* can be said to undermine the prevalent racial value judgment given expression in such racial categories as European/non-European and civilized/non-civilized. It subverts the racist assumption of superiority. For lack of a better concept, Swift's satirical vision of the Europeans' place in the world could be described as sceptical moral anthropology. With their representations of geographical space, Defoe and Swift begin an intertextual dialogue, which quickly turns into a real public debate on the relationships of Europe with the rest of the world. And this debate is so reminiscent of the English tradition of academic and political debates that it is hard to escape the suspicion that the connection between the two works is purely accidental.[42] One cannot but be struck by the light that Defoe's and Swift's fictions shed on each other.

Robinson Crusoe and *Gulliver's Travels* are not just imitations of genuine travel accounts. Their cartographic representations of space and their geographies lend their stories a similar character and at first contact invite a similar type of reading. This is as it should be. For Swift's work, this means that Lilliputians, Brobdingnagians, Laputans, Yahoos and even Houyhnhnms must be considered foreigners, like the Japanese of Book III, and like Crusoe's Friday, before they can be recognized as satirical mirror images of Gulliver and Europeans in general. With their stories, the two authors push the fictional lies to their logical

41 *A New Voyage round the World* distinguishes between a larger number of societies but the basic opposition between whites and non-whites remains.

42 The comparison of the geographies of *Robinson Crusoe* and *Gulliver's Travels* sheds light on what Claude Rawson has recently called the "teasing relationship" between their works (*TLS*, Sept. 10, 2004, 3). It can be taken as evidence that Swift was probably more intimately acquainted with Defoe's fiction than is usually assumed, even though he may not have owned a single copy of the latter's works and may have disliked the man.

consequences. Aesthetically, both writers use types of fictional deception. Defoe, the inventor of geographic realism, presents his novels as true stories taking place in the real world.[43] In *Robinson Crusoe* already, his mapping of the hero's voyages and island is supposed to fulfil fictionally the function attributed to maps and cartographic descriptions from the fifteenth to the nineteenth century "of opening up and closing a territory" and of "naming [places] to indicate possession and imminent settlement", as William Boelhower points out.[44] Thus it is that the island that Crusoe discovers becomes "Crusoe's island": an island all the more real for being fictional. Defoe's fictional geography is the basis of his realism. It invites the reader to share the proposed illusion of the real and, by extension, the narrator's vision of life. In Swift's parody of contemporary travel accounts, the maps and depictions of Gulliver's islands add another touch to his criticism of the unreliable, delusive, and sometimes fanciful, character of contemporary cartography as already noted about his geography of the sea. It debunks the idea that mapping a land gives a right to possess it. It can also be said to deride Defoe's use of geographical realism in fiction. Fully aware that "it is … perfectly possible for a map to be false, not only because it is an inaccurate or falsified diagram, but also because its role in a complex message is such as to undercut, rather than underline, the truth of the representation",[45] the author proposes a tale in which, unlike Crusoe's, the only thing that Gulliver can ultimately boast of to his readers as the proof that the lands and islands he has discovered exist are his cartographic descriptions of them and the maps, all of them "falsified diagrams". No one else can corroborate his findings, as he is the only survivor of his outward-bound voyages. As a result, the territories he has opened up cannot be "closed", in other words claimed. But it does not matter, as the author of the *Travels* himself, whose portrait faces the title page, is that of a person who does not exist and is a fiction, like Crusoe.

43 W. R. Owens, "Introduction", *The Life and Strange Surprizing Adventures of Robinson Crusoe*, 22.

44 William Boelhower, "Inventing America: A Model of Cartographic Semiosis", *Word and Image* 4 (1988): 483, 481–82.

45 Stephen Bann, "The Truth in Mapping", 498.

With their maps and cartographic representations of maritime and land space, Defoe's first novel and Swift's satire were pioneering works of fiction. Their treatment of space marked a turning point in the evolution of realism as representational form. Geography, the maps of the world and of individual countries or specific regions, became an inescapable reference, even in works of fiction that took liberties with these in the name of fantasy like Swift's. After them, the cartographic and geographical conception of space informed such different works of fiction as Henry Fielding's and Tobias Smollett's novels, Laurence Sterne's *Sentimental Journey*, and even the setting of his *Tristram Shandy*. For his part, Oliver Goldsmith plays variations on Defoe's theme of his hero's return by land from the Far East in *Farther Adventures*. In his *Citizen of the World*, Lien Chi Altangi, the Chinese visitor travels from China to England by a parallel route to Crusoe's while, in a second journey, Altangi's son reaches England by way of Persia and a route that reminds one of yet another traveller and visitor in Europe, Usbek, in Montesquieu's *Lettres Persanes*. It is hardly an exaggeration to say that without this geographic sense of space the novel would not have been possible. The weakness of Defoe's and Swift's cartographic representations of the world, heirs as they were to the era of the great explorations and its cartography, was the deficiency of their topography. The remedy would come later in the century from a different source.

3. Moll Flanders, Joseph Andrews and A Few Other Travellers on the English Network of Post Roads

Early modern fiction did not only relate daring overseas voyages to remote exotic horizons. It also portrayed characters travelling closer to their homes, around their own country. The scenes change but not their geographical representation. The scale is smaller, the spaces better known, even familiar so that they are no longer subjects of wonder or debate. The descriptions are more detailed, yet their nature is still cartographic. The same interplay between the development of fiction and contemporary map-making can be traced as in the case of fictions about overseas voyages but the reference is to another type of geographical representation. What immediately draws the attention is an interest in roads and, to a lesser extent, towns and their plans. In the same way as Swift's and Defoe's geographies conjure up the routes of voyagers and explorers across seas, continents and islands when they sketch out Gulliver's and Crusoe's movements, the novelists' geographies focus on the roads and streets followed by their characters in their travels. From the beginning of the eighteenth century, writers resolutely turned the backs on the world of romance with its "elsewhere, unmarked and unchartered":[1] on what a nursery rhyme calls the dream of getting to Babylon and back by candlelight. Their aim was to portray credible characters journeying on real roads connecting existing villages and towns, pushing further than the picaresque novel the semblance of geographical authenticity. The heroine's journeys in *Moll Flanders* (1722) and the account of the hero's return home from London in *Joseph Andrews* (1742) are eloquent illustrations. This representation of movements on actual communication routes, which often takes into consideration the mileage and time needed to travel between places, was something new. It followed in the wake of a recent concern with roads. Roads were becoming of prime political and economic importance on the national level. Their traffic was beginning to rival that of sea and

1 Douglas Chambers, *The Reinvention of the World: English Writing 1650–1750*, 21.

river routes. In fact, what the age witnessed was the birth of a new conception of roads as a network facilitating the movements of people on land and goods on the regional, national and continental levels. This network structured the environment in a novel way and was codified in a special type of atlas and maps: the road book and road map. Road geography was another manifestation of "utilitarian geography", the dominant conception in the eighteenth century.[2]

Before the Restoration, maps of the British Isles and European countries in general showed towns, villages, rivers, clusters of hills or mountains, perhaps a few estates but they seldom showed existing roads. This was still the case in the splendid maps of Europe in Johan Blaeu's *Atlas Major* of 1665. Blaeu's map of part of Kent (fig. 3.1) shows villages that seem unconnected with one another. After the Restoration, roads were added. As the age became increasingly concerned with "systems of mensuration",[3] the new maps were based on more accurate surveys of the land and calculation of distances. One device used to obtain an exact mileage for the distances between towns was the "wheel dimensurator" or "foot-wheel" of a circumference of half a pole (21/4 yds. or 2.615 meters) (fig. 3.2). In England, John Ogilby was the first to produce road maps with the new measurements in his atlas, *Britannia* (1675).[4] The most innovative mapmaker of his time, he also records distances between towns using the statute mile of 1760 yards. His ribbon maps are presented like strips in a way reminiscent of Peutinger's Roman tables. They each unwind a central double line showing a segment of the course followed by an important road. The double and single dots between the two lines indicate the distances in miles and furlongs respectively. Two continuous central lines mean that the land on both sides of the road was enclosed by hedges. Dotted lines on either or both sides show common land. Rows and clusters of tiny rectangles along the road symbolize villages and towns. Their arrangement presents the rough plan of the place. A single house stands for a manor. The other symbols are isolated churches, bridges, groupings of hills, trees for parks and forests and thicker or thinner wavy lines for rivers.

2 Paul Claval, *La Pensée géographique: Introduction à son histoire*, 13.

3 Douglas Chambers, *The Reinvention of the World: English Writing 1650–1750*, 22.

4 John Ogilby. *Britannia, London 1675, with an Introduction by Dr. J.B. Harley* (Amsterdam: Theatrum Orbis Terrarum, 1970).

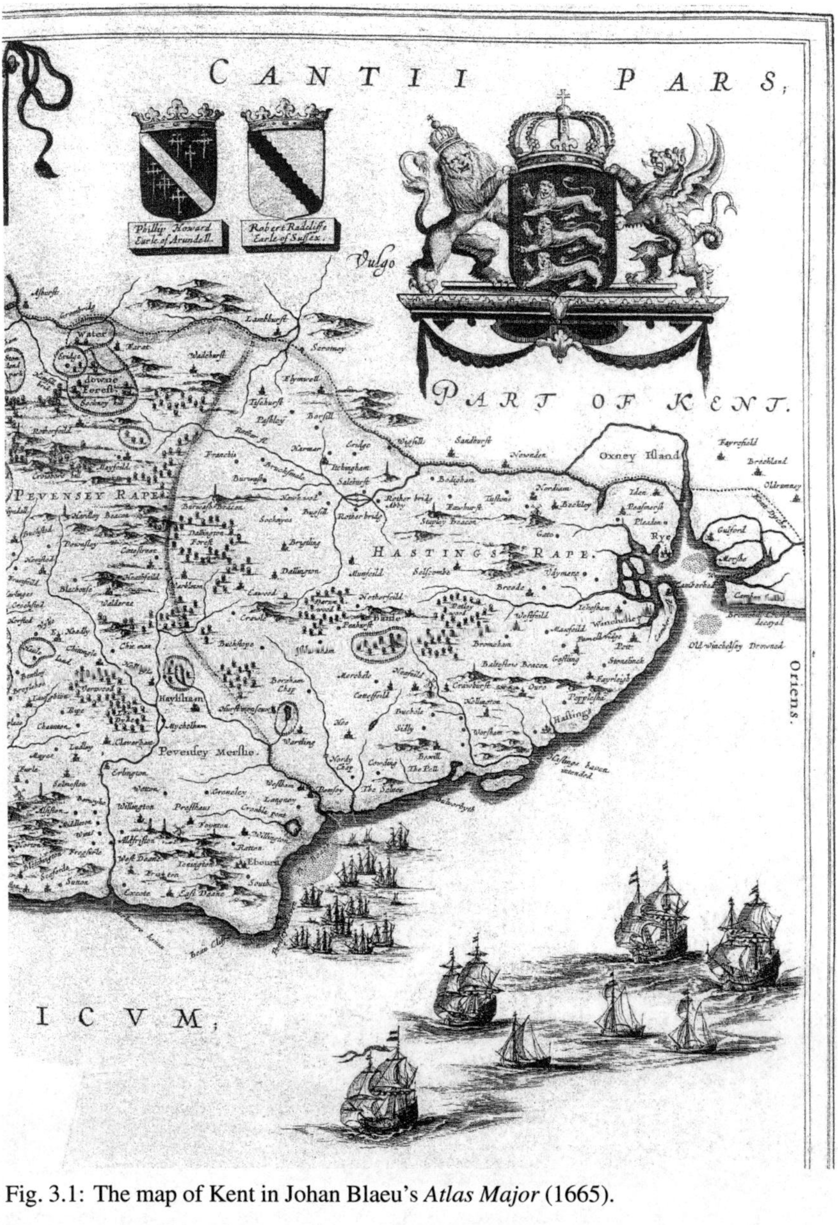

Fig. 3.1: The map of Kent in Johan Blaeu's *Atlas Major* (1665).

Fig. 3.2: Ogilby's atlas, *Britannia*: central part of the frontispiece. It presents the activity on the roads and the land surveyors' instruments. On the right-hand side, the "wheel dimensurator". The picture is signed "W. Hollar fecit 1675", the geographer and "scenographer" referred to at the beginning of the previous chapter.

Fig. 3.3: The network of the roads of England and Wales in Olgilby's atlas *Britannia*.

The maps are accompanied by descriptions of the villages, towns and a few topographical features that the roads passed through. A general map is included, which made it possible for the first time in history to visualize all the main roads of England as a network (fig. 3.3). The fact that Ogilby's atlas was reprinted for 25 years and frequently imitated confirms the demand for such maps. Atlases and travellers' road books in the same pattern became such popular publications that over thirty titles and reprints have been listed just for the period 1675–1730, two of them by geographers that have already been mentioned in connection with sea charts, Herman Moll and John Senex.[5] By the time Defoe and Fielding launched their characters on their travels across England, the proliferation of road books was a further proof that roads had become of major interest to a large public and it was inevitable that this interest should in some way or other find expression in fiction. Unlike world geography, road geography was at once accepted as presenting indisputable measured reality.

The kinship between the presentations of roads in atlases and the representation of road journeys in the new fiction is even closer than in the case of world maps and overseas travels. First of all, this is because road maps and their accompanying comments in road books provide information on the routes available to travel around the British Isles and Defoe's and Fielding's choice of itineraries for their characters sometimes seems to have been inspired as much by maps as by personal experience. The fact that it is not unusual for Defoe's characters to discuss their routes to a specific destination in terms reminiscent of someone consulting a road map or road signs at least suggests it. This is how two of his characters, Colonel Jacque in the eponymous novel and a companion, deliberate upon which road to take on leaving Newmarket in order to rejoin the Northern Road to Scotland:

> The next Question was, what Road we should take; here were four ways before us, and we were alike Strangers to them all; first on the right Hand, and at about a little Mile from the Town, a great Road went off to *St. Edmund's-Bury*; strait on, but

5 Herbert George Fordham, *The Road-Books and Itineraries of Great Britain, 1570 to 1850: A Catalogue with an Introduction and a Bibliography* (Cambridge: Cambridge University Press, 1924). R. A. Skelton, *County Atlases of the British Isles, 1579–1850: A Bibliography* (London: Carta Press, 1970).

> enclining afterwards to the right, lay the great Road to *Barton Mills*, and *Thetford*, and so to *Norwich*; and full before us lay a great Road also to *Brandon* and *Lynn*, and on the left, lay a less Road to the City of *Ely*, and into the Fens.[6]

But it is mainly the parallelism of the representations of itineraries in road books and fiction that is noteworthy. Geographical features are presented in the same sequential order. This takes a visual form in maps as they unfold their ribbons. Where comments are annexed to the maps, they take the form of a succession of brief descriptions and, in travel accounts, the sequential order of these features is part of the basic structure of the narrative. The procedure is linear, both spatially and chronologically. It defines a territory. In road books and fiction, the resulting territorialization is the same. Defoe's and Fielding's heroes, not only Moll Flanders and Joseph Andrews, are depicted travelling from London or some other place in various directions, making use of the network of roads that geographers depicted as extending to a region, the whole national space or even, in some cases to parts of continental Europe. Needless to say, the range of these journeys is more impressive in Defoe than in Fielding. In its variety, Defoe's road geography is indeed reminiscent of that of his overseas voyages.

More unexpected is yet another similarity. Cartographers and fiction writers give pride of place to the same main roads. In his description of the English network of "highways", as important roads were commonly called, Ogilby distinguishes between four types. The most important are the "Direct Independents". They connect London with other harbours—Bristol, Harwich, Holyhead, Milford Haven and St. David's, namely the gates to overseas destinations, the Continent and Ireland—and with provincial centres—Colchester, Chester, Cambridge, and Gloucester. When a regular service of transport and post was available, these "Direct Independents" were also referred to as "Post Roads". The roads of the second category, the "Direct Dependents", branch off from the "Direct Independents" and lead from London to still other, more remote centres like Carlisle or Norwich. Then come the "Principal Cross Roads", connecting provincial centres with each other, and finally the

6 *The History and Remarkable Life of the Truly Honourable Col. Jacque*, 102.

"Accidentals", consisting of shorter branches. For their part, Defoe and Fielding make their characters travel essentially on roads of the first two types, namely the ancestors of today's A30, A4, A5, A10, A11, A12 radiating westwards, northwards and eastwards from London, the very roads that the geographer describes as of national and economic importance. The suggestion is no longer of single ventures across oceans but of repetitive movements between towns and residences. The kinship between atlases and fiction invites the readers to follow the characters' itineraries on maps, in the same way as, in modern editions of *Moll Flanders*, they do with the heroine's movements around London on reconstituted maps of the city.[7] The following table lists Moll's fifteen road journeys, those of a few other characters in the same novel and that of the six central characters in *Joseph Andrews*. Opposite, their routes as described in Ogilby.

Moll's Road Journeys	
1. COLCHESTER – LONDON (65)[8] Moll leaves Colchester to settle in London with her first husband.	On the Yarmouth–London "Post Road", which joins the London–Harwich "Direct Independent" road passing through Colchester (49 miles and 7 stages). "From *LONDON* to *Rumford* and so to *Burntwood* 16 Miles; to *Ingerstone*, then to *Chelmsford* and so to *Witham* 18. to *Keldon* and thence to *Colchester* 12. to *Ipswich* 16. *Saxmundham* 16. *Beckles* 16 and *Yarmouth* 10."[9]
2. LONDON – OXFORD – NORTH-AMPTON – LONDON (67) A tour that Moll undertakes with her second husband.	London–Islip–Aberystwyth "Direct Independent" road, branching out shortly before Islip at Wheatley-Bridge to Oxford (55 miles).

7 See the section "Maps" in *Moll Flanders*, Penguin Classics (London: Penguin, 1989).

8 *The Fortunes and Misfortunes of the Famous Moll Flanders*, ed. Liz Bellamy, *The Novels of Daniel Defoe* (London: Pickering and Chatto, 2009). All references given in the text are to this edition.

9 The distances in the quotations are indicated in "Post Miles", which were rough computations different from those of Ogilby's new "Dimensuration". The spelling of the towns and villages has been modernised. (*) indicates that the road is not described in *Britannia*.

	Oxford–Islip–Towcester–Northampton on "Principal Cross Roads" or "Accidentals"* (approx. 50 miles). Return by way of Stony-Stratford, first on an "Accidental",* then from Stony-Stratford to London on the Holyhead–Chester–London "Post Road", a "Direct Independent" (53 miles).
3. MILFORD HAVEN – LONDON (99–100) On her return from Virginia, Moll lands at Milford Haven after a rough crossing of the Atlantic. Her journey takes three weeks.	"Direct Dependent" or "Accidental"* road to Haverfordwest, then the St. David's–London "Direct Independent" road passing through Kidwelly, Llanelthy (Llanelly), Swansea, Aberavon, Cowbridge, Corntown, Landoff, Cardiff, Newport, Mitcheldean, Gloucester, Barnesley, Fairford, Lechlade, Faringdon, Abingdon, Dorchester, Henley, Maidenhead, Colnbrook, Hounslow, Brentford: approx. 250 miles.
4. LONDON – BRISTOL – BATH (100) Moll goes to Bristol to look after her cargo of goods from Virginia and decides to settle in Bath.	"Post Road" and "Direct Independent". The London–Bristol road comprised 7 stages: "From *LONDON* to *Hounslow* 10 Miles; to *Maidenhead* 16. to *Reading* 12. to *Newbury* 16. *Marlborough* 15. *Chippenham* 15. and *BRISTOL* 20." Bristol–Bath on a "Direct Dependent"* (approx. 130 miles).
5. BATH – BRISTOL and back (101) Her several business trips.	"Principal Cross Road"* (approx. 13 miles).
6. BATH – SHEPTON and back (105) Moll fetches her sick Bath friend from Shepton.	From Bath to Shepton: 15 miles on "Accidental Roads".*
7. BATH – BRISTOL – GLOUCESTER – BATH (106) A "journey of pleasure", like Journey 2, which Moll undertakes with her Bath friend.	"Principal Cross Road"* to Bristol, then the Bristol–Worcester "Principal Cross Road" to Gloucester and back (approx. 90 miles).[10]

10 This "Principal Cross Road" and one of the Upton inns play an important role in another novel referred to below, Henry Fielding's *Tom Jones*, Books IX–XI.

8. BATH – HAMMERSMITH (109) Moll leaves Bath to settle closer to the father of her child.	Bath–Chippenham on a "Direct Dependent" and Chippenham–Hammersmith on the Bristol–London "Post Road" or "Direct Independent".
9. LONDON – LANCASHIRE (126) Moll leaves London hoping to live more cheaply in the country and marries an impecunious robber husband.	The London–Chester–Holyhead "Post Road" or "Direct Independent" to Stone-in-Cheshire, then from Darlaston-Bridge near Stone, the London–Carlisle "Post Road" and "Direct Dependent" to Warrington and eventually an "Accidental" to Liverpool (approx. 190 miles and 12 stages). "From *LONDON* to *Barnet* 10 Miles; to *St. Albans* 10. *Dunstable* 10. *Brickhill* 7. *Towcester* 13. *Daventry* 10. *COVENTRY* 14. *Colshill* 8. *LICHFIELD* 12. *Stone* 16. *Nantwich* 16. *CHESTER* 14." From Stone to Warrington by Bruerton Green.
10. LIVERPOOL AREA (126–27) Day trips to a country seat and from there to a village in the neighbourhood of Liverpool (198).	Unspecified "Accidentals"*.
11. LIVERPOOL – BLACK ROCK – CHESTER – DUNSTABLE – LONDON (128, 137, 140) Moll's return to London from Lancashire. Her Lancashire husband accompanies her to Dunstable. From there, she proceeds alone to London.	By boat to Black Rock, and then by hired coach to Chester on an "Accidental".* From Chester, the London–Chester–Holyhead "Post Road" and "Direct Independent". See Journey 9 above (182 miles and 12 stages).
12. LONDON – STONE-IN-CHESHIRE (154) Moll leaves London to make her banker friend and next husband believe that she is still living in the country.	The same London–Chester–Holyhead "Post Road" and "Direct Independent" as in Journey 9 above (141 miles).
13. STONE – STONY-STRATFORD – LITTLE BRICKHILL – DUNSTABLE – LONDON (154–161) Moll is supposed to be travelling back from Lancashire to meet her next	The London–Chester–Holyhead "Post Road" and "Direct Independent". See Journey 9 above. In fact, this is the return journey from Journey 12.

husband at Stony-Stratford, and she eventually marries him at the next stopping place, Little Brickhill.	
14. LONDON – DUNSTABLE and back (184–85) Moll, now a thief, leaves London because she fears she may be arrested.	The same London–Chester–Holyhead "Post Road" and "Direct Independent" as in Journey 9.
15. LONDON – STURBRIDGE (Cambridge) – NEWMARKET – BURY ST. EDMUNDS – IPSWICH – HARWICH – IPSWICH (by boat)– COLCHESTER – LONDON (216–20) Moll's thieving expedition around East Anglia. In *Colonel Jacque*, Defoe's eponymous hero on his way to Edinburgh follows the same route as Moll for the first part of his journey.	Northern "Post Road" and "Direct Independent" London–Newcastle to Royston. Then, the Royston–Norwich "Principal Depending Branch" of the "Post Road", a "Direct Dependent" first to Cambridge, then to Newmarket and Bury St. Edmund. From Bury to Stowmarket on an "Accidental,"* and from there on the King's Lynn–Harwich "Principal Cross Road" to Ipswich and Harwich. Return to London on the London–Yarmouth "Post Road", which joins the Harwich–London "Direct Independent" before entering Colchester. The town on a river whose name Moll does not remember (341) must be Stratford-Street on the Ouse.
* * *	* * *
The Road Journeys of Other Characters in Moll Flanders	
16. LONDON – COLCHESTER (65) Moll's first husband travels back to Colchester to attend his brother's wedding.	See the route of Journey 1 above.
17. BATH – LONDON (107) Moll's Bath friend travels three times back to London on business or to see his family.	See the route of Journeys 4 and 8 above.
18. LONDON – LITTLE BRICKHILL (154) Moll's banker friend and husband-to-be travels from London to Little Brickhill to meet her and travels back to London with her as his new spouse.	See the route of Journey 14 above.

19. LONDON – BARNET – TOTERIDGE (198) Three victims of Moll's in London travel back home to Barnet	London–Chester–Holyhead "Post Road" and "Direct Independent".
20. LONDON – STURBRIDGE (Cambridge) (216) Mention must be made of the gang of thieves with whom Moll travels to Sturbridge Fair.	Northern "Post Road" London–Newcastle to Royston. Then, the Royston–Norwich "Direct Dependent" or "Principal Depending Branch" of the Post Road.
21. HOLLAND – HARWICH – LONDON (217–18) The business route of the Dutch traveller whose portmanteau Moll manages to swipe under the nose of his servants in Harwich.	Crossing of the sea and then the London–Harwich "Direct Independent".
22. LONDON – LANCASHIRE – LONDON (140–48) Moll and her banker friend exchange letters before their marriage	The correspondence travels on the same London–Chester–Holyhead "Post Road" and "Direct Independent" as in Journey 9.
* * *	* * *
The Six Main Characters' Journey in "Joseph Andrews"	
LONDON – SOMERSET Joseph, Fanny, Parson Adams, Mrs Slipslop, Peter Pounce and Lady Booby travel back to Somerset by the same route. Two of them, Fanny and Parson Adams, in fact begin the journey by travelling in the opposite direction, towards London.	London–Land's End "Post Road" and "Direct Independent" entering Somerset shortly before reaching Yeovil (11 stages). The road leaves London from Cornhill, Cheapside, St. Paul's Churchyard, Ludgate-Hill, Fleetstreet, Strand, Haymarket and Piccadilly. *"Kensington (4 Miles); Hammersmith (2); Brantford (8); Staines (7); Bagshot (8); Hartley-Row (8); Basingstoke (8); Andover (16); Salisbury (15); Shaftsbury (18); Sherborn (12)".*

The table draws attention to the two writers' distinctive handling of road and travel geography. Wherever his characters go, Defoe names and locates with geographical precision the places on the map of England or of the world. In his accounts of their road journeys, as in the case of their overseas voyages, he is principally concerned with points of departure and destinations. It is commonplace in criticism to say that his narrative method focuses on the important episodes of his characters' lives. These episodes are generally connected with the places where the heroes or heroines reside for a while. Accounts of journeys and voyages are mere transitions between one scene of their lives and another, unless something special happens during the trip, a delay or accident of the kind that befalls the main character at sea in *Robinson Crusoe*. But journeys are seldom eventful in Defoe, and they never are in *Moll Flanders*, which can be said to present the geographical component in Defoe's narrative art in its baldest form. The reports of her journeys seldom exceed a few lines. The author is content to describe her routes. The following are two examples: one describes Moll's first journey to Bristol and Bath after her return from America (Journey 4 in the above list):

> The looking after my Cargo of Goods soon after oblig'd me to take a Journey to *Bristol*, and during my attendance upon that Affair, I took the Diversion of going to the *Bath* … (100)

In the second, Defoe relates her removal from Bath to Hammersmith and the circumstances in which it takes place (Journey 8). Moll explains that her Bath friend had provided lodgings for her at Hammersmith and she simply adds that she

> accordingly hir'd a Coach on purpose, and taking my Child and a Wet-Nurse to Tend and Suckle it, and a Maid Servant with me, away I went for *London*. (109)

Elliptical statements like "I took the Diversion of going to the *Bath*" and "away I went for *London*" reduce the journeys to the role of narrative and spatial links between episodes. They link up stopping places much like the lines showing roads on a ribbon map. Such brief accounts present the character's journeys once more as trajectories. Thus, at first sight, Defoe's representations of his characters' movements look at best sketchy. As will be seen below, however, it would be a mistake to conclude from their brevity that these minimal accounts are negligible.

Fielding's representation of geographical space is more concrete and detailed than Defoe's and closer to that found in the descriptions accompanying road maps. On the other hand, he never names a village or small town, only perhaps a county, so that his geography is hazy. This absence of geographical names allows the writer greater fictional freedom in his handling of space. If, in *Joseph Andrews*, the author asks the reader to excuse him for not being "particular as to the way [his characters] took" to reach Booby Hall,[11] he nevertheless presents their journey from London to a village somewhere in Somerset in such a way that it can be identified with only two possible routes described in contemporary road books. Joseph either had to take the Bath road like Moll or, which is more likely, to follow the London – Land's End trunk road described in the table above, which leaves London from Cornhill and Cheapside and enters Somerset after approximately 120 miles.[12] Ogilby seldom mentions particular inns but he does "The Lion" at Staines. Now, "The Lion" is the very name Fielding gives the "House of Hospitality well known to the Western Traveller" where Joseph first stops after leaving London (Book I, Ch. xi, 49), which confirms that the second hypothesis is the correct one (fig. 3.4).[13] The second inn Joseph reaches on horseback 20 miles further (Book I, Ch. xi. 50) must then be at Hartley-Row, a hamlet now part of Hartley Wintney, "a place of some entertainment" at the time, according to Ogilby.[14] Ogilby's description of the attraction of the place, however vague, is interesting, because it may explain why the owner of the horse that Joseph rides to this place should be waiting there for it. Two miles further, the hero is set upon by highwaymen (Book I, Ch. xii, 51). This would situate the robbery somewhere before Joseph reaches the small village of Merrard Green.

11 *Joseph Andrews*, ed. Martin C. Battestin, The Wesleyan Edition of the Works of Henry Fielding (Oxford: Clarendon, 1967), Book III, Ch. 2, 191–92. All references given in the text are to this edition.

12 I am aware that studies using a biographical approach sometimes identify Lady Booby's seat with East Stour, a village in Dorset between Shaftesbury and Sherborne, which significantly shortens the journey (Pat Rogers, *Henry Fielding: A Biography*, London, Paul Elek, 1979, 122). Yet it must be remembered that, when Fielding attended the assizes on the Western Circuit, which he did regularly for many years, he took the same route and went much further (see Martin C. Battestin, *Henry Fielding*, London, Routledge, 1989, 272–73, 283, 308 309–10, 316, 351–52, 357, 373, 375, 383, 391, 396–97, 409, 421).

13 Ogilby, *Britannia*, 50.

14 Ogilby, *Britannia*, 50.

The stagecoach that rescues him drops him at an inn "four miles" away (Book I, Ch. xii, 52). The place must be relatively small as the coachman only stops briefly for some refreshment. The distance suggests Newnham rather than Hook.[15] All this is more precise than one might have expected. *Tom Jones* (1749) presents the same kind of cartographic outlining of the hero's route to London in the central part of the novel. Tracing his itinerary incidentally shows that, on two stretches, between Bristol and Gloucester and from Stony-Stratford to London, Tom's journey recalls Moll Flanders' movements.[16]

In Fielding, the experience of the road itself is more central than in Defoe. His narrative dwells on his characters' actual awareness of distances covered and of the pace at which they proceed between stopovers. The novelist also makes several references to the topography of the land along the road. Like the cartographer, he occasionally indicates the crossing of rivers and a few ascents and descents of the road without being specific about the breadth of some or the declivity of others. However he never dwells on the relief of the land in the way Bunyan does in *The Pilgrim's Progress*. The stopovers at inns and these topographical features are mentioned in the sequential order of their presentation on maps and in road guides and punctuate the narrative. Fielding also refers to places in the immediate neighbourhood of the road, a space equivalent to that shown on both sides of the roads in strip maps. Once it is a squire's mansion (Book II, Ch. iii, 102) and, on another occasion, an out-of-the-way house where Joseph and his two companions spend one night (Book III, Ch. ii–iv, 194–229). As for the Downs where Parson Adams rescues Fanny from her ravisher (Book II, Ch. ix, 136–37), they could correspond to one of the clusters of molehills beyond Shaftesbury on Ogilby's map (fig. 3.5). Fielding may take liberties with

15 In this case, the sign of the inn, "The Dragon", would seem to be an emblem of the character of the innkeeper's shrewish wife and not a reference to an actual inn, the dragon not being a common name for inns or alehouses except in connection with saint George (Book I, Ch. xiv, 61).

16 See the map included in the Norton edition of *Tom Jones,* ed. Sheridan Baker (New York: Norton, 1973), 769. *Tom Jones* uses much the same strategy as *Joseph Andrews*. Its geography becomes explicit once the hero leaves Somerset. When Tom is sent away, he wants to go to Bristol but ends up taking his first night abode at an alehouse on the road to Gloucester. The context makes it clear that he and his guide got lost on what is at best a road that Ogilby describes as an "Accidental" (Book VII, Ch. 10).

the naming of places and the exact location of topographical features in his description of the route followed by his characters, but the parallelism between the narrative and strip maps is nevertheless remarkable.

Before pursuing the examination of the roads travelled by Moll Flanders and Joseph Andrews, a few words should be added about the historical interest of contemporary atlases in general and what they can contribute to a better understanding of the novelists' geographical descriptions of England at the time. They are first a useful reminder that the country was very different from today and that the places named have changed radically since. As noted in the introduction, seventeenth- and eighteenth-century maps and their accompanying descriptions draw attention to aspects of the characters' travels that now all too easily escape the reader's attention for the simple reason that the authors refer to them implicitly, assuming that their contemporaries could and would have supplied the information from their own experience—novelists always do this to some extent, some more, some less. Maps and road books are consequently useful in helping fill in the geographical background with valuable information. This is particularly the case with Defoe, whose geography relies on the reader's familiarity with the names of places and indications of routes to create a sense of the land. To that extent, it must be admitted that, with the exception of the description of Crusoe's island, the geographical component in the realism of his novels is at best allusive.

Leafing through Ogilby's atlas, the reader has a clearer idea of the kind of places where Defoe's and Fielding's characters stop along the way. The cartographer usually refers to these stopping places as "towns". One must not be misled by the word and by their names. Except for London, most "towns" were little more than villages by modern standards, and it was these larger or smaller villages that the trunk roads went through and connected with London and with each other. What importance these towns had, they owed either to the postal service that served them, to a crossroads, to a local industry, or to the regular fairs or markets held in them. The following account of a segment of the London – Chester – Holyhead road on which Moll travels four times is a telling illustration:

> … entering *Bedfordshire*, an enclos'd Way leads you at 34 Miles in *Dunstable*, from one Dun a notorious Robber so call'd, anciently *Magoninium* and *Magintum*;

'tis seated on a Hill in a chalky dry Ground, well-frequented, full of Inns for Accomodation, and has a good Market on *Wednesdays* for Corn and Cattle, and is Noted for good Larks, having a cross Road in it, the Left to *Leighton* and *OXFORD*, and the Right to *Cambridg*.

At 34'3. leaving the Town your road gently ascends to 35'4. where a steep Descent carries you down to *Chalk-hill*, thence (if in Winter) a dirty Way leads you to *Hockley*, alias *Hockley in the hole*, (from *Hoc* i.e. *Lutum*) long and scattering, consisting chiefly of Inns and publick Houses; and here the road to *Woburn* branches out on the Right. Hence the *Watlingstreet* way brings you to the entrance into *Buckinghamsh*. at 42'4. 9 Furlongs farther entering *Little Brickil* 3 Furlongs long, a Post-town and of good Reception. Then gently descending at 45'5. you pass the *Lowsel* a Branch of the *Owse*, and enter *Fenny-Stretford* 2 Furlongs in Length, and mostly Inns for Reception of Travellers. Thence by *Rickley*-wood on the Left, over *Bunch*-hill and *Harford*-bridg a straight Road leads you at 52'6. into *Stony Stretford*, seated on the *Owse*, anciently *Lactoradum*, a large Town containing two Parish-Churches, well-accomodated with Inns, has Five Fairs yearly, and a great Market on *Fridays*, their chief Imployment being Bone-lace weaving; At the End of the Town, having past the river, you enter Old *Stratford* in *Northamptonshire*, affording some good Inns, both together making a Mile in Extent.[17] (See fig. 3.6)

None of the places mentioned in this passage are of any significance today and none are mentioned in Defoe's *Tour through the Whole Island of Great Britain*. The only one that still occasionally figures in traveller's guides is Dunstable. On the other hand, Little Brickhill, a post town in Ogilby and in Moll's lifetime, has dwindled into such insignificance that it has disappeared from general road maps. Further north, where today the A5 becomes the A51 and in the area where one would expect Ogilby to name Birmingham and Stoke-on-Trent, he and Defoe mention Stone. At the time, Stone-in-Cheshire may not have been much larger than Birmingham and Stoke-on-Trent but it was the town north of which the road to the North-West branched out, hence its greater importance. Still further north, Defoe mentions Liverpool, which Ogilby and *Britannia Depicta, or Ogilby Improv'd* of 1736 ignore, although they mention nearby Warrington: an oversight explained by the fact that Liverpool had not yet become the prosperous harbour it was destined to be.[18]

17 Ogilby, *Britannia*, 22.

18 John Owen's *Britannia Depicta or Ogilby Improved* of 1736 still does not mention Liverpool in its "Table of Cities, Towns, etc.", though the relevant maps show two "Accidentals" leading to it, one from Warrington and the other from Newton-in-the-Willows (92).

With its repetitive focus on distances to travel between towns and on inns, the above quotation from Ogilby also recalls Joseph Andrews' slow progress from inn to alehouse and alehouse to inn on the London – Land's End road. Fielding's hazy geography makes no mention of towns, although his hero had to pass through three sizeable ones, Andover, Salisbury and Shaftsbury. As seen above, he mentions a famous inn, "The Lion", but he does not bother to associate this inn with Staines. The reason why Fielding hardly ever indicates that his hero stops in towns can only have been that he expected his readers to understand implicitly that Joseph's inns and alehouses were nearly always situated in towns or villages along the road. He only hints at the fact. Indeed, it is only in "towns" that Joseph and Parson Adams could have access to a surgeon (Book I, Ch. xii, 55), a constable (Book I, Ch. xvi, 70), a Justice of the Peace (Book II, Ch. xi, 145), or to a Parson Trulliber (Book II, Ch. xiv, 162–68), when they needed their help. Atlases fill in this background at the same time as they remind modern readers that in none of the larger towns the houses stretched for more than three or four furlongs (600–800 meters) along the road: the length of an average street in modern cities. It would be a mistake to imagine that Joseph's inns or alehouses are situated at lonely crossroads.

Atlases and road books do not only tell us that towns and villages were much smaller than they are today but also that England as a whole was sparsely populated in the eighteenth century, another fact not foregrounded by fiction because it would have been obvious to readers, yet perfectly discernable in the handling of space. Furthermore, the country was less cultivated than it is today, and the agrarian revolution that was soon to hasten the pace at which common farmland was privatized and enclosed by hedges—historians speak of the enclosure movement—had not yet remoulded the landscape as we know it. The farther away from London one got, the more thinly inhabited the countryside was, which of course influenced communication and people's movements. Ogilby's mapping of the London – Land's End road and Fielding's account of Joseph's journey illustrate this. The ribbon road map and the accompanying description show that, at first, on leaving London, travellers could hardly journey four or five miles without passing through a hamlet, village or town and, in the novel, the inns and alehouses at which Joseph and the other characters stop cannot be very distant from one another (fig. 3.4). From the moment one entered Wiltshire (fig. 3.5),

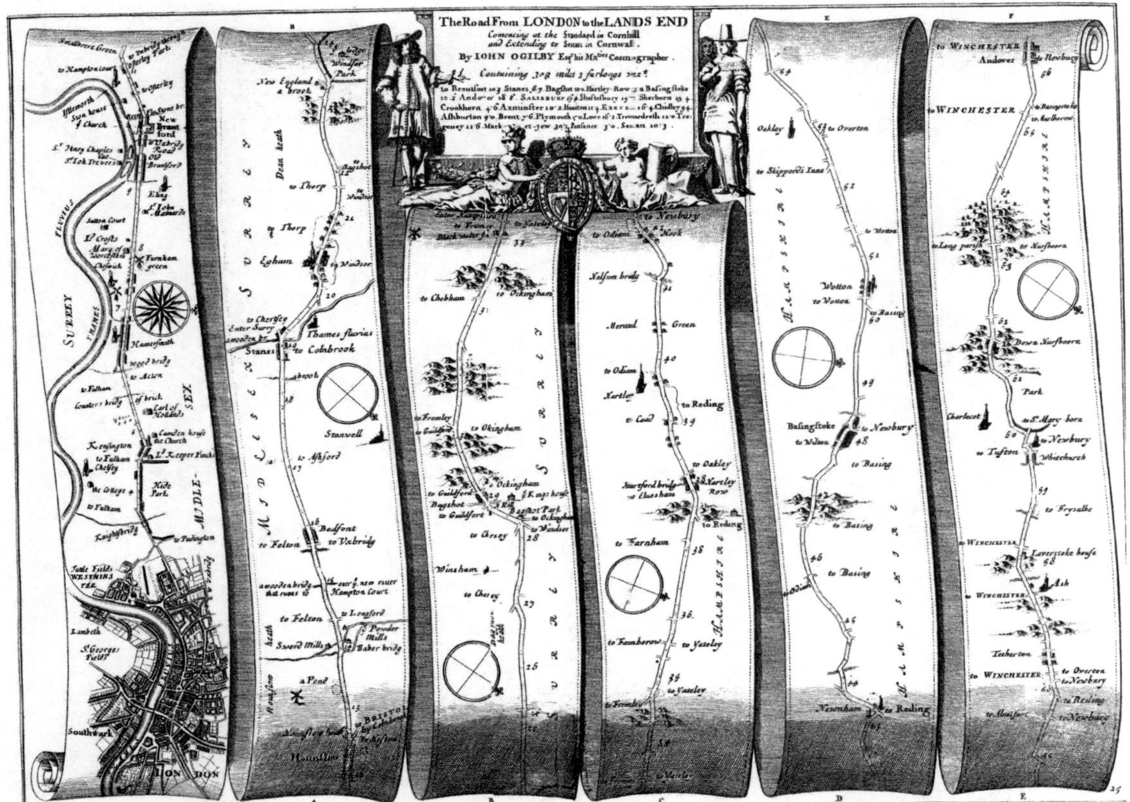

Fig. 3.4: The London–Winchester section of the London–Lanc's End strip map in Ogilby's atlas *Britannia*.

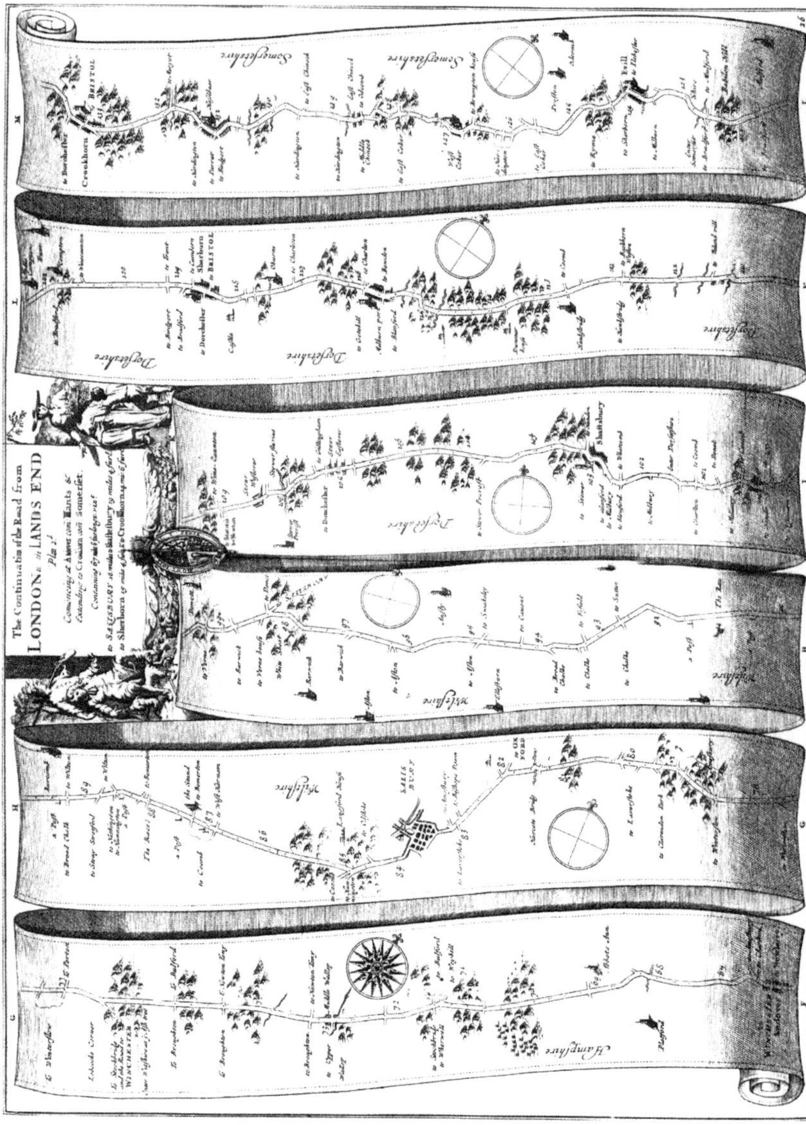

Fig. 3.5: The Winchester–Crookhorn section of the London–Land's End ribbon roadmap in Ogilby's atlas, *Britannia*.

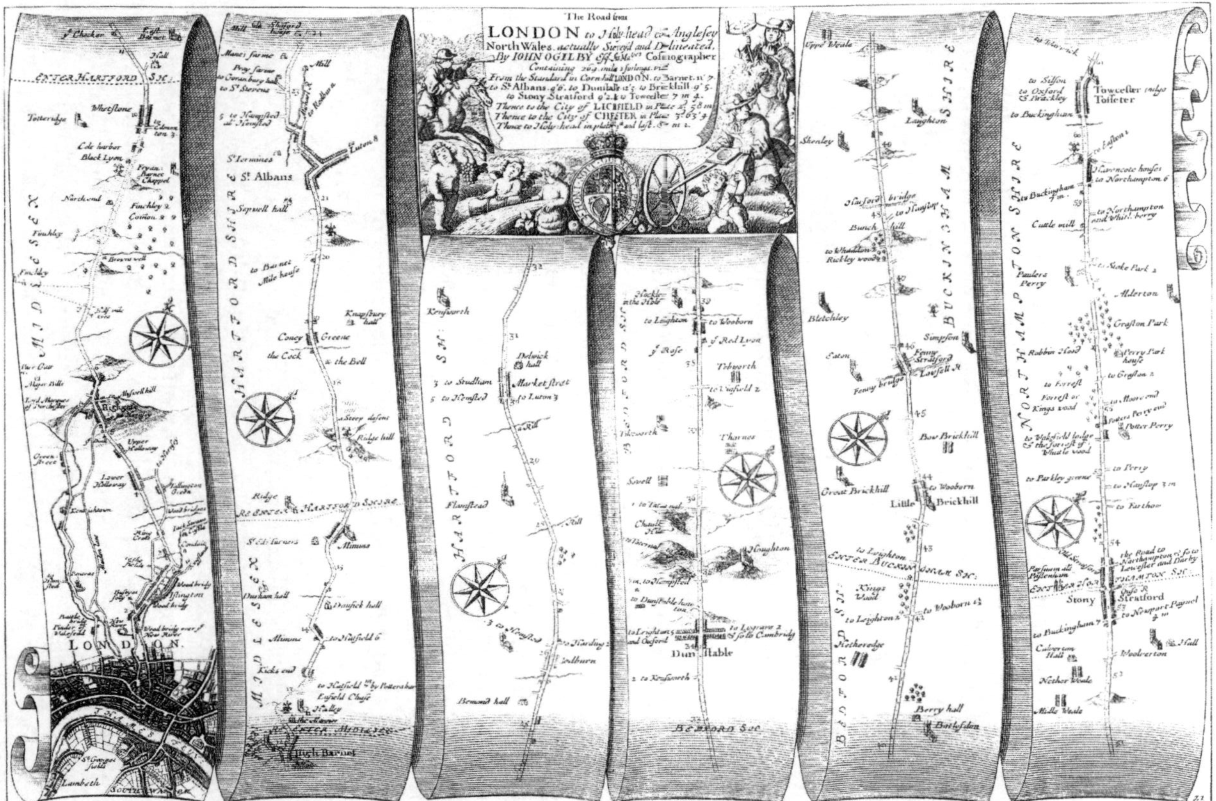

Fig. 3.6: The London–Tow Chester section of the London–Holyhead strip map in Ogilby's atlas, *Britannia.*

villages became less frequent and the countryside undulated more, so that travellers sometimes had to walk or ride a dozen miles without passing even a hamlet. In the later episodes of Joseph's journey, the novel takes into account this change in the lie of the land and density of population, in particular when Joseph, Parson Adams and Fanny, after trying to escape from people they take for robbers, are unable to reach the next inn at nightfall. They precipitately come down a hill, cross a river and walk across meadows to the already mentioned lonely house of Mr. Wilson's, where they eventually find shelter (Book III, Ch. ii, 193–94). The very next day, they are victims of the dogs of a hunt during their midday siesta on the road (Book III, Ch. vi, 236–43). Two such incidents following each other so closely could only take place in a little inhabited area where farmland was not yet enclosed. Far away from London, the traffic would inevitably be lighter. Such information helps convey the impression that the country looked larger and that, to people living in the Western Counties, the Midlands or further north, London seemed more remote than it does today.

There are even cases in which the confrontation of atlases and road books with the fiction shows how the new realism dictated its geographical choices to the authors. Ogilby's depiction of the section of the road beyond Dunstable quoted above thus explains why Moll's Lancashire husband is able to ply his trade as a robber so close to London. It mentions a stretch of road crossing a hilly and little inhabited area, a fact confirmed by the map (fig. 3.6). Incidentally, there is also some reason to believe that the very idea of the incident of the Dunstable Hill robbery (Journey 13, 247–49) was the legend of the robber, Dun, to which Ogilby refers. The descriptions of Stony-Stratford and Little Brickhill in the same quotation read like an ironical comment on the circumstances of Moll and her banker friend's wedding. His intention is to meet her and marry her at Stony-Stratford, a town with two respectable churches, says the road book. Marrying in a church in such a town would have given a desirable official character to the ceremony. Unfortunately, owing to the hazards in the communications and transports of the time, they only manage to meet and marry at an inn at the next stopping place on the way to London, Little Brickhill, a mere post town, which leaves the impression that the wedding takes place at a kind of Gretna Green (154). Still, Little Brickhill is better than the unspecified place in Lancashire where Moll and her robber husband had been hastily married only a short time before (128). Another of Ogilby's descriptions, the description of Hammersmith, helps to understand why Moll's Bath friend chooses lodgings for her there. Hammersmith was, according to the atlas, a good upper-middle class town near London.[19] The choice underlines the man's liking for his mistress and his regard for her as the gentlewoman she claims to be. Contemporary readers, or at least some of them, could not but have been aware of such overtones, which would escape us today were it not for such information found in contemporary guidebooks. It brings out the graphic character of the geography, enlivening the narrative of road journeys, as does the greater attention that Fielding pays to topography and the small size of towns.

From the perspective of the modern historian, English roads at the time were not much better than fifty years earlier when John Bunyan in *The Pilgrim's Progress* dwelt on their swampy ditches, ruts and quagmires. Of this, the geographer's classification of roads, of course, said little, but the fiction writers could not but take such things into consideration in their evocation of their characters' mobility. Bad conditions were most to be deplored during the winter season and in rainy weather but roads with little traffic, even important ones, might become overgrown with grass during the better seasons, so that the main track nearly disappeared making it difficult for the traveller to find his way. As late as 1771, Henry Mackenzie describes such a situation in *The Man of Feeling*. At one point on the way home, his main character, he says,

> entered a hollow part of the road, which winded between the surrounding banks [of a river], and seamed the sward in different lines, as the choice of travellers had directed them to tread it. It seemed to be little frequented now, for some of those had partly recovered their former verdure.[20]

Moreover, to the traveller, the geographer's distinction between the four categories of roads was far from being always as evident as the road books made it. For a short distance, a main "highway" might turn into a mere track and, at a crossroads, a secondary road branching off might appear wider and more frequented than the main highway, as Fielding's Parson Adams once learns at his own expense (Book II, Ch. vii, 130).

19 Ogilby, *Britannia*, 19.

20 Henry Mackenzie, *The Man of Feeling*, Oxford World Classics (Oxford: Oxford University Press, 2001), 63.

To obviate such confusions in the absence of road signs, the more reliable road books usually provided lists of wrong turnings to be avoided. Most novels of the time mention such difficulties. It must be remembered that milestones and signposts only began to appear in the second half of the century, with the improvements brought about by the increase of turnpikes, the toll roads of the time.[21]

What is more difficult to account for is that Defoe, Fielding and their successors did not make more of these physical hurdles than they did.[22] Not even in *Tom Jones* does the author dwell on the shameful conditions of roads, although the novel relates movements of its characters on some principal and secondary roads during the winter season. The fiction of the first half of the eighteenth century, like road books, puts the emphasis exclusively on motion, distances travelled and communication between places. This relative silence on the poor quality of the roads cannot be explained by the development of more efficient forms of transport either. The only explanation would seem to be that the age developed a new, more dynamic perception of movement through space, an impression reinforced by the preference of the two writers for the depiction of long-distance journeys on Ogilby's "Direct" thoroughfares, namely principal roads to and from London. Their geography is of people who have no time to waste. Enhanced, or at least more widespread mobility did not mean that the pace at which people and the novelists' characters travelled was much faster than in the previous century. Except for couriers on horseback, the average speed remained anything between five and ten miles an hour—ten miles an hour in favourable conditions and on fairly flat roads for those able to afford suitable means of conveyance. Progress was so slow that it was necessary for travellers to stop for refreshments, meals and overnight stops, which further lengthened the journeys. Travelling on land had changed less since Roman times than it was to do between the nineteenth and twentieth centuries.

21 John Brewer, *The Pleasures of the Imagination: English Culture in the Eighteenth Century*, 640.

22 Defoe and Fielding ignored the poor quality of roads in their fiction but not in their lives. In 1697, Defoe could still complain of their sad condition and "many places unpassable" in "An Essay upon Projects", *Social Reform*, ed. W.R. Owens, in *Political and Economic Writings of Daniel Defoe,* vol. 8 (London: Pickering and Chatto, 2000), 55–56.

With the preceding remarks on the quality of roads and the travellers' experiences on them, we have already moved from the discussion of the road as space, the space enclosed between the double lines on ribbon maps, to the subject of the road as "un lieu où se déroule le vécu" to use Henri Lefèbvre's phrase once again. This space where living takes place is unlike any other. As seen above, there is the national space or territory that their network covers, the larger space in which roads inscribe their trajectories and fictional writers chose the direction of their characters' movements. On both sides of the roads, there are enclosures, areas of common land, houses, mansions and churches, all of them spaces produced by individuals or communities with a view to permanent exploitation—the way these spaces were used reminds one of Crusoe's appropriation of his island. Roads are different. They are spaces produced by one community for other communities as much as for itself. The community takes the space away, subtracts it, from the land on either side so as to open it up to all and sundry without restriction. Roads are a space of motion within the larger space of the nation or region, a space of transit and communication, a space of traffic of people, animals and goods, a space in which ideas travel. The travellers' utilization of its space is limited in time to their presence on them. As for the travellers' experience of the road, it cannot be compared to that of navigators and explorers like Gulliver or Crusoe. The latter were pathfinders facing elementary problems of survival in their explorations. The situation for road users is completely different. They are a digest of English society and are conscious that they cannot escape its rules and the presence of their fellow men and women in its narrow space.

In most cases, the roads the characters follow are not simple highways: they are also "Post Roads", according to Ogilby, namely roads offering a regular service for conveying mail and an equally regular service of public transport. The best known was the service of stagecoaches for the middle classes. It connected London with the main towns and harbours of the country. Slower and less comfortable were the wagons, covered vehicles for the regular transport of commodities and poor passengers. Finally, there were the post horses, horses that could be hired at one inn and exchanged at another to travel fast. Such regular services on so many roads could not have existed but for an equally regular demand on the part of large numbers of passengers travelling long distances within the national territory. Beside public transport there were, of course, private vehicles and horses. *Moll Flanders* and *Joseph Andrews* mention coaches and six—coaches drawn by six horses—coaches and four, chaises, chariots, hired coaches. At some time or other in her life, Moll avails herself of nearly all the existing types of vehicles, which confirms the impression that travelling is in her blood. She favours the faster and more comfortable means of locomotion. When no particular vehicle is specified, she must be assumed to travel by stagecoach, which perfectly suits her as a would-be gentlewoman. She mentions hired coaches three times (Journeys 6, 8 and 11). On one occasion she has the pleasure of travelling in a private coach and four (Journey 10) and once even, like a duchess or Fielding's Lady Booby, in a coach and six with two footmen, a coachman and a postilion or driver (Journey 2). Once she takes advantage of a chance coach returning to London from a special trip (Journey 13, 154) and, another time, she stoops to get into a wagon to return to London (Journey 15, 220). Among Moll's acquaintances, her Bath friend is reported to return from Shepton to Bath (Journey 6, 105) in a litter, and one of her suitors comes to meet her in a chariot (Journey 18, 154), the same type of conveyance used by Lady Booby's steward, Peter Pounce (Book III, Ch. xii, 271). To the list of vehicles mentioned by the two writers must be added the horses that gentlemen and their footmen rode. Finally there were the pedestrians. Joseph Andrews, a footman who has lost his job, as well as Parson Adams, his mentor, and his beloved Fanny, must be content to walk much of their way back to Somerset. The references in the two novels to the many types of public and private transport as well as to men on horseback and pedestrians bring to life what road books and atlases imply with the notion of post roads. They conjure up a picture of constant traffic by day and often by night.

Fiction illustrates what the popularity of road books and almanacs, often providing the same information, hint at: an increasing number of people from all stations of life used the "highways" and other important roads. *Joseph Andrews* lays stress on the different social classes. The nobility, the gentry, the professional classes and the servant classes are represented in the novel. Travelling on the same road to Dorset or Somerset as Joseph does, there are Lady Booby, another gentleman in a coach and six with his servants and dogs (Book I, Ch. xvi, 74), an earl's grand-daughter (Book II, Ch. v, 123), Lady Booby's estate steward, Peter Pounce (Book I, Ch. x, 47; Book III, Ch. xii, 270–71), an exciseman (Book I, Ch. xvi, 75), several clergymen and lawyers, a bookseller (Book I, xvii, 79), chambermaids, down to footmen like the hero and other menial servants like Fanny. Defoe is more concerned with the social and economic status of his essentially middleclass travellers. Among them, there are a banker, a Dutch businessman, a young man travelling between London and his family at Colchester (Moll's first husband), and a man estranged from his wife (Moll's Bath friend). Above all, there is Moll who, unlike Richardson's Pamela, is many women in succession and plays many roles from a married woman to an independent woman and from an expatriated businesswoman returning from America to a thief on the prowl or on the run.[23] *Tom Jones* adds companies of soldiers (Book VII, Ch. xi), a party of gypsies (Book XII, Ch. xii) and a beggar (Book XII, Ch. iv). Nor must the people working in the transport services be forgotten: the coachmen and postilions who drove these people around. Finally, there was the occasional highwayman. Defoe's and Fielding's characters illustrate the great variety of early eighteenth-century travellers crowding the roads and the inns along the way. What seems to have struck them was the number of women among them. Fiction conveys the impression of new categories of people, and even of a whole society on the move.

On the move, but not at the same pace. The general impression may be one of easy movements between places, but the speed at which people journeyed depended on their mode of locomotion, which itself depended on their social and economic status. The cost of transport does not pose Defoe's characters any problem and the author of *Moll Flanders* has little to say on the subject. The focus on routes gives the impression of rapid journeys. Fielding is more sensitive to differences. In *Joseph Andrews*, the rich are seen to travel faster than the other classes. Fastest are the travellers on horseback. Shortly after leaving London, Joseph is first overtaken by a footman on one of his master's horses. Lady Booby leaves London last but, with her coach and six, she catches up with all the other members of her household. Fielding writes that she "overtook the other Travellers as they entered the Parish", their destination (Book IV, Ch. i, 277). She probably does not spend more

23 "moins double que multiple", writes Serge Soupel in "Moll Flanders: Duplicité, mensonges, prudence," *Etudes anglaises* 51 (1998):19.

than one or two nights at inns on her way to her country residence. Her steward's chariot, which overtakes Joseph earlier, is clearly slower, drawn as it is by one or two horses, yet it is faster than the stagecoach in which Mrs Slipslop, the chambermaid, travels.[24] At one point, Parson Adams is described as walking at such a brisk pace that the stagecoach starting from the inn shortly after him does not catch up with him (Book II, Ch. vii, 130). Wagons travelled at much the same pace as pedestrians. They mostly offered a possibility of saving one's legs and the collective security of numbers on long distances. Travellers on foot, who plodded on from inn to inn, were the slowest of all. The slower the journey, the more exposed the traveller was to all sorts of accidents and adventures.

In Defoe and Fielding, the characters travelling on post roads or at least "Direct Independents", "Direct Dependents" or "Cross Roads" are mostly on the move for practical social and economic reasons: business, the need to find new lodgings, a change of status or job. There is little here of the "involuntary, unguided drifting" that Kevin Cope finds in later eighteenth-century fiction.[25] Journeys on main roads were obviously not conducive to this kind of leisurely roaming. Roads lead to precise destinations. Necessity is a frequent cause of movements. It concerns the heroes and heroines. They are driven into taking the road by an insecure social status and fear of destitution. Joseph returns home after losing his job as a footman in London. Resilient Moll is in constant search for new suitable places to settle down. At first, she travels to look after her financial affairs. As said above, to check on her cargo from Virginia she takes the same road to Bristol that Swift's Gulliver must take in order to join the crew of *The Antelope* when he leaves on his first voyage. She goes to Lancashire hoping to live more cheaply there than in London (Journey 9). A summer fair motivates her journey to Cambridge (Journey 15): she has been told that Sturbridge can offer profitable trade to the thief that she has become. It is worth noting that the same fate affects Defoe's and Fielding's other main characters,

24 In the last third of the century, stagecoaches could travel from four to seven miles per hour and from 70 to up to 120 miles a day. At the time Defoe's or Fielding's characters travelled, the speed was still much slower.

25 Kevin Cope, "Drifting: Unguided or Involuntary Motion and the Enlightenment Sense of Direction", Papers of the Third International Paul-Gabriel Boucé Conference, Paris, 2006.

Tom Jones, Captain Singleton, Colonel Jacque, and even Roxana. The recurrence of the situation highlights a sensitive social phenomenon: numerous travellers on the roads were economic victims of society in one way or another. As for the majority of the secondary characters, the narrative implies or shows that they travel for family or professional reasons. This is particularly true of the passengers in stagecoaches. Moll's first husband goes back to Colchester to attend his brother's wedding (Journey 16) and the Dutch gentleman in Journey 21 is obviously on a business trip to England. Parson Adams initially sets off for London to find a publisher for his sermons. An interesting case is that of Lady Booby's. Fielding's novel implies that she regularly moves with her whole household between her London and country residences, as upper-class families usually did. Not all the journeys are motivated by necessity or obligations however. There are a few exceptions in *Moll Flanders*. When the heroine goes to Bristol to dispose of her cargo, she makes a detour to Bath to enjoy the social life there (Journey 4). Then twice she sets out, once with her second husband and on another occasion with her Bath friend, on tours to visit the sights of the country (Journeys 2 and 7). *Moll Flanders* is one of the first expressions in fiction, if not the first, of this very modern pastime, touring for pleasure and visiting the sights of a country. For the middle classes, it was a new custom favoured by the greater facility to travel around the country and fostered by the curiosity aroused by descriptions of these sights in a new generation of travellers' guides like Defoe's own *Tour through the Whole Island of Great Britain* (1724–1726). The Grand Tour of the sons of the nobility had awakened an interest in the monuments and beauty of one's own country.

Fiction shows no interest in the traffic of commodities, which was at least as important as that of people, particularly in the surroundings of London, as the same *Tour* testifies. On the other hand, it refers to the quick circulation by road of opinions and news, whether by letter, word of mouth or books, as well as of rumours. Moll and her banker friend exchange news by letters during their courtship (140–48). In *Joseph Andrews*, the reader is given to understand that Whitefield's Methodist ideas, the views of the Deists and Parson Adams' latitudinarian ideas have come to Parson Barnabas' shocked ears by road (Book I, Ch. xvii, 80-84). In *Tom Jones*, rumours that Bonnie Prince Charlie has landed in the north and is marching upon London with an army of Jacobite rebels twice reach an inn where one of the main characters makes a stop (Book VII, Ch. xi, 367; Book XI, Ch. ii, 576–78).[26] Fiction also depicts roads as channels through which morals, manners and new customs spread and bring about changes. The custom of travelling for pleasure or taking a holiday at Bath has been mentioned, and such holidays are not quite innocent. Moll easily finds a friend and protector at the spa town, a married man estranged from his wife and they become lovers during a sightseeing tour (Journey 7). Mrs Slipslop, Lady Booby's chambermaid, makes a second attempt—the first was in London—to seduce Joseph when her stagecoach overtakes the young man on the road (Book II, Ch. iii, 101). The fact that soon after she offers him to ride with her in the chaise sent for her by her mistress makes it clear that she intends to make a third attempt (Book II, Ch. xiii, 159). Mention must finally be made of Betty, the chambermaid at "The Dragon" "who is daily liable to the Solicitations of Lovers of all Complexions" (Book I, Ch. xviii, 86). With this account of Betty's life, Fielding hints that roads play a role in the transmission of venereal diseases.[27] In both writers, the road is sexualised. *Moll Flanders* and *Joseph Andrews* furthermore concur that, with the development of an efficient road network and better means of transport, the evils and immorality of city life are more likely to contaminate the country than the country to restore moral health to the city. Lady Booby and Mrs Slipslop bring back with them to the country their indecorous infatuation with young Joseph. In Defoe, thieves, Moll among them, leave a London deserted by the wealthy during the summer to comb the surrounding counties. On the other hand, Parson Adams fails to find a publisher for his sermons in London—in fact, he and his sermons never get there—and, ironically, Joseph is sent back from the city precisely because he has proved too virtuous. It is true that, with its highwaymen and sheep stealers, criminality is not absent in the countryside. *Joseph Andrews* even has a squire invite Joseph, Parson Adams and Fanny to his hall to take advantage of Fanny's innocence (Book III, Ch. vii–viii, 244–55). To these extreme cases of

26 In the second case, the rumour reaches an inn that has been identified: "The Bull's Head" at Meriden on the London – Holyhead Post road and "Direct Independent", north of Coventry and before the old road to Birmingham branches out to the left (574).

27 Defoe in *A Journal of the Plague Year* also mentions the role of roads in the spreading of diseases.

criminality or depraved behaviour must be added the more frequent instances of unscrupulous innkeepers and their staff, who "always proportioned [their] Respect to the Appearance of a Traveller" (Book II, Ch. v, 119) and cheated or at least overcharged guileless guests. These evils are to be largely imputed to the development of the road network itself and to the ease with which people move around the country. In *Moll Flanders* and *Joseph Andrews*, the road can be accused of contributing to a lowering of moral standards.

For Defoe, and even more so for Fielding, the road is not simply a zone of motion and transit: it is also what anthropologists call a contact zone, where people from different social and geographical backgrounds rub elbows. This, however, does not mean that it is a space favourable to the development of companionable relationships. The two novelists agree on this point: bent as their characters are on their own business, they have no greater desire to mix than in town or in the country. For them, the road as contact zone does not draw the different social classes closer to each other. In this sense, the novel qualifies Michael McKeon's contention that, in the eighteenth century, "the several social ranks experienced an unprecedented degree of public intermixing". As an aspect of "the out-of-doors", roads are by no means a "realm of commoners—the people, the many, the mob—" as are the streets of London.[28] In their great majority, road users in fiction belong to the upper and middle classes. The rich travel in their own coaches and have their meals in the privacy of separate apartments at inns, imposing on the road what Michael McKeon calls "their sense of indoor privacy, privilege, and enclosure."[29] The middle classes journey in stagecoaches and have their meals in other private rooms, and the lower classes take advantage of wagons, or go on foot, and warm themselves at the kitchen fires of alehouses. Contacts with local people are not always easy either. The road with its nodal points, the inns and alehouses, and the narrow spaces of stagecoaches or wagons, are at most the site of exchanges of news and ideas, besides the spreading of morals and diseases. True, they make contacts between so many people from different backgrounds inevitable, but they are often the sites of frictions.

The new genre turns these situations to good account. Not that all contacts are unfriendly. Moll and her husband are once escorted by an

28 Michael McKeon, *The Secret History of Domesticity*, 166.
29 Michael McKeon, *The Secret History of Domesticity*, 20.

obliging innkeeper, his son and three stout fellows on a stretch of road where they might encounter highwaymen (Journey 13, 161), and she once finds a country bumpkin ready to let her ride with him to the next town (Journey 15, 219). Joseph Andrews is once offered a ride to the next inn by a footman of his acquaintance (Book I, Ch. xi–xii, 50–51) and he, Parson Adams and Fanny are hospitably entertained by Mr Wilson (Book III, Ch. ii–v, 194–229). Frequently, however, quarrels arise and even brawls start among travellers or between travellers and local people. The conversation about religion between Parson Adams and Parson Barabas turns into a heated argument between people of the same profession but holding different views. Sometimes travellers from the upper classes or considering themselves such resent direct contact with their inferiors. Joseph is twice the innocent cause of quarrels between the passengers of a stagecoach: once when, victim of a robbery during the night, he is found lying naked and half-conscious in a ditch and some passengers in the stagecoach refuse to let him ride with them in the coach to the next inn (Book I, Ch. xii, 52–53), and once when a young woman passenger delays the departure of a coach because she refuses to travel in the company of a footman:

> The Coachman began now to grow importunate with his Passengers, whose Entrance into the Coach was retarded by Miss *Grave-airs* insisting, against the Remonstrances of all the rest, that she would not admit a Footman into the Coach; for poor *Joseph* was too lame to mount a Horse. (Book II, Ch. v, 122–23)

Miss Grave-airs is of the opinion that Joseph, a footman, should await a wagon if he needs a conveyance. (Book II, Ch. v, 123). Fielding's third-person narrative and humorous approach lends itself well to the evocation of such difficult contacts on the road. The narrow social space of the road with its annexes, inns, alehouses and stagecoaches inevitably stir up antagonisms.

Rooted as they were in the travellers' minds and in their way of life, social barriers were in fact as present on the road as anywhere else in society, as Defoe and Fielding well knew. Yet there were moments when these barriers fell because roads were spaces of relative anonymity, and this anonymity became a source of freedom from social constraints, not simply of freedom of movement. Defoe's fiction often refers to this situation. The role of roads, inns and by extension spas in the case of love affairs between people of different social status has been pointed out. There are other instances. On the road, Moll and her second husband, who have hired a coach and six, pass easily for a lady and her lord (Journey 2). In a stagecoach or at an inn, as long as they can dress suitably and pay their fare, Moll or Mrs Slipslop claim to be gentlewomen. Unless travellers come across acquaintances, as Joseph does, they could be whoever they liked on the road. Better still, on the road, people could conceal their identity. At Dunstable in Journey 14, Moll puts up at an inn where, in Journey 11, she stayed with her Lancashire husband, who is no longer her husband, but she pretends to be waiting for him to join her there, although she is in fact a thief, who has run away from London to escape the police. When she wants to change her status from widow to wife (Journeys 12 and 13) or wife to widow (Journey 9), she uses her road journeys to cover her tracks. The freedom of movement equally favours other sorts of deceptions. Moll's use of the postal service on the London – Chester road has already been referred to. The Penny Post introduced shortly before 1680 and the fact that coffee houses in London and inns in the country were mail collection points enables her to conceal her present situation from her banker friend. Like him, she lives in London. But to make him believe that she is still a single woman and quietly waiting in Lancashire for his marriage proposal, she lets him send his letters to her there and has them forwarded back to her in London. As space of fleeting contacts between characters, the road, with its dubious anonymity, looks like an extension of the sprawling modern city into the country.

Defoe's and Fielding's depictions of roads as man-produced collective spaces of enforced sociability are both similar and complementary and this is not a surprise, considering that they depict the same historical background. But their characters are as unlike each other as one would expect from their respective authors. Moll Flanders' and Joseph Andrews' experiences on the road demonstrate this. The authors can be said to play variations on the motif of the road journey. Defoe's representations of his characters' travels in *Moll Flanders* may be so brief as to appear insignificant, mere links between episodes, but what precedes has made it clear that they acquire quite another dimension through repetition. In fact, they tell us quite a lot about Moll and the other characters. The frequency of Moll's movements, her fifteen journeys, and the accumulation of miles she travels by coach, along with her four crossings of the Atlantic, prove such a central experience in her life that the reader remembers her as a character constantly on the move, whether out of necessity—twelve out of the fifteen journeys— or simply because she can never resist the idea of an excursion (Journeys 2, 4 and 7). The fact that many secondary characters are also depicted as going places suggests that Moll, with her restlessness, is representative of her time rather than unusual. Presented as a seemingly endless succession of movements between locations which are always different, Moll's journeys form a pattern that may be interpreted as a consequence of her having no roots and no place she can call her home. Captain Singleton and Colonel Jacque are also in search of roots and a home that would give them a social status. As for Roxana, she becomes a rootless kept woman when her husband deserts her and deprives her of what should have been her and her children's home. Lack of roots and homelessness are definitely causes of instability. Then, these minimal accounts of the characters' movements suggest that they have little time for contacts and are in a hurry to get to their destinations. Even when going places in the company of other people, they tend to keep to themselves as the great individualists that they are. Their routes are their trajectories of life. Their travels also spotlight the importance of money as a factor of mobility and autonomy, the autonomy that Crusoe was struggling to achieve in completely different circumstances. "With Money in the Pocket", Moll says, "one is at home any where" (154). Money is presented as more important than social status and "le paraître" as more important than "l'être" in claiming or holding a share of the space of the world.

For Fielding, the novelist of short stages and of the road with its margins as represented on ribbon maps, the road is emphatically a contact zone in which travellers from various backgrounds cannot always avoid contact with each other and with local people. His characters are gregarious. Travelling is a less individualistic undertaking. The title characters have companions who they can count on. The representation of road geography moves in the direction of demographic and, consequently, social geography. At the same time, his vision of society is more conservative than Defoe's. His characters cannot escape their rank and class on the road. If they eventually do so, it is only by the trick of the happy ending. In Defoe, the road, like the sea, is the site of the main characters' often frantic or anguished quests for a better, more acceptable social and economic status. In Fielding, the road is first synonymous with the loss of a stable position and then with a search for rehabilitation or an equivalent social or economic solution. Another characteristic trait in Fielding is that the road leads the character back home. Again, unlike Defoe's characters, Joseph and Tom Jones are not rootless. Their homes are one in Dorset or Somerset and the other in Somerset. For Joseph, it is the village where the object of his affection, Fanny, and of course his mentor, Parson Adams, reside. For Tom, it is Squire Allworthy's manor house, where he grew up, and its neighbourhood.[30] The name of the place, Paradise Hall, says everything. Home, for them as for most people, is the traveller's safe haven. It is where one goes for rest, help or comfort. But home is first and foremost, in the perspective of the present study, a space and place that characters and readers can locate in the British Isles and on a map like any known place. The characters' travels are still depicted as trajectories: trajectories of people following each other on the same roads, even pursuing each other. In *Joseph Andrews*, the journey of the main characters assumes the form of snakes and ladders, whereas, in *Tom Jones*, the suggestion is of a hunt with the males on the females' trails, an added humorous touch to the author's mock epic.

After Defoe and Fielding, the novel hesitated between the geographical precision of one and the more allusive topography of the other. At the same time, journeys and voyages were more subtly integrated in broader patterns of fiction, with the result that they often became less central compared to other aspects of eighteenth-century life. Yet Smollett's *Roderick Random* (1748) and Henry Mackenzie's *The Man of Feeling* (1768) offer memorable pages on road journeys. One contains the most vivid fictional account of a journey by wagon and on foot—Roderick and his friend Strap's journey from Newcastle to London, twenty-one stages and 276 miles 4 furlongs according to Ogilby. The other, Henry Mackenzie's short novel, proposes, besides the description of a stretch of road overgrown with grass quoted above, one more portrayal of a typical group of middleclass stagecoach passengers and of the way the title character produces his space in the vehicle.

30 Paradise Hall is usually identified with Fielding's birthplace, Sharpham Park, south-east of Glastonbury in Somerset (Henry Fielding, *The History of Tom Jones, a Foundling*, ed. Fredson Bowers, The Wesleyan Edition of the Works of Henry Fielding (Oxford: Clarendon, 1974), n. 42.

> The company in the stage-coach consisted of a grocer and his wife, who were going to pay a visit to some of their country friends; a young officer, who took this way of marching to quarters; a middle-aged gentlewoman, who had been hired as house-keeper to some family in the country; and an elderly well-looking man, with a remarkable old-fashioned periwig.
>
> Harley, upon entering, discovered but one vacant seat, next the grocer's wife, which, from his natural shyness of temper, he made no scruple to occupy, however aware that riding backwards always disagreed with him.[31]

"Made no scruple to occupy" is an odd phrase to describe a man taking the remaining seat in a coach but it is one that draws attention to the fact that even such a trivial action is a claim on space, an individual's production of space, be it only for the duration of a journey.

In Smollett and later novelists, precise geographical references to places are treated in such a way as to highlight the fact that, whether in their movements on land or at sea, the characters cross one another's paths, often several times, in their search for a place in the world, in other words in their production of a viable social and economic space for themselves. This handling of space evokes the picture of the world as a competitive arena, where ambitions and interests collide to the advantage or detriment of those involved. The road and the sea with their nodal points, towns, vehicles and ships, are increasingly treated as contact zones and settings of social life. From the trajectory of individual lives, fiction had moved to the representation of complex collisions between numerous individuals.

One novel of the period, Smollett's *The Expedition of Humphry Clinker* (1771) occupies a special place in the development of the fictional representation of geographical space. Its epistolary form, inspired by contemporary collections of travel letters "disseminating information on geography, history, customs, and manners",[32] can indeed be said to relate it to the famous correspondences of Montesquieu's Persian visitor in Paris in the *Lettres persanes* and of Goldsmith's Chinese in England in *The Citizen of the World*. True, Smollett's novel is in a lower key, dealing as is does with British not international themes, yet it is no less interesting for that. Its letters relate a tour through the Island

31 Henry Mackenzie, *The Man of Feeling*, 57.

32 Byron Gassman, "The Economy of *Humphry Clinker*", *Tobias Smollett: Bicentennial Essays Presented to Lewis M. Knapp*, ed. G. S. Rousseau and P.-G. Boucé (New York: Oxford University Press, 1971), 156.

of Great Britain—the title of Defoe's guide could have been a fitting one for the fiction. The journey is undertaken by a group of his Majesty's subjects. Their letters are sent to friends. They do not only report their adventures and convey their travel impressions but are also full of personal concerns. In this sense, they are more reminiscent of another type of epistolary exchange, of which Samuel Richardson's *Pamela* is the prototype. The conflation of these two types of letters in *Humphry Clinker* draws attention to the fact that, whether by road or sea, the characters' letters travel between senders and recipients, as Moll Flanders' letters to her banker friend do in Defoe's novel. In fact, the letters travel as much as the characters. In *Humphry Clinker*, the epistolary framework presents the itinerary of the characters' journey. The list of places in the letter headings gives the outline of the tour: Gloucester, Clifton Hot Well near Bristol, Bath, London, then up the east coast to Harrogate, Scarborough, Newcastle upon Tyne, Berwick-upon-Tweed, Edinburgh, Cameron, Glasgow and down the west coast to Carlyle and through the west Midlands. The last letters are vaguer about the exact places where the characters stop than those of the journey up to Scotland. The contents of the travellers' letters complete the geographical information, adding precisions on the houses, inns or streets where they take lodgings at each stage. Every important spot that Matthew Bramble from South Wales, his sister, Tabitha Bramble, their nephew, Jerry Melford, and their niece, Lydia, visit is the point of departure towards the next planned destination. Presented this way, the geographical description of their route bears similarities to Defoe's highlighting of itineraries. Incidentally, many of the roads the characters follow are once more the very highways made familiar to contemporaries by maps and atlases and taken up by previous fictional travellers. The Gloucester – Bristol, Bristol – Bath, Bath – Marlborough – London routes had been followed by Moll Flanders. After leaving London, the Bramble party takes the same Northern Road as Roderick Random, but in the opposite direction. Like Colonel Jacque and his companion before them, they do not follow it straight to Edinburgh but make "detours",[33] usually, in their case, to visit spas. On the stretch between Wetherby and Borough Bridge, they turn left to spend a few days at Harrogate, a spa famous for its ferruginous water, and then leave

33 Tobias Smollett, *The Expedition of Humphry Clinker* (New York: W.W. Norton, 1983), 263. Further references given in the text are to this edition.

the Northern Road again to go to Scarborough after stopping at York. On their way back from Scotland along the West Coast, they must be understood to follow a stretch of the Stone-in-Cheshire – Carlyle "Direct Dependent" that branched off to Manchester, which they visit.

The epistolary form enables the writer to combine Defoe's cartographic pattern with descriptions of the roads and scenes of what happens on the way or at stopovers in the style of Fielding's. In the south, Matthew Bramble complains of the dust, of the press of post-chaises and other vehicles "besides the troops of fine gentlemen that take to the high-way, to shew their horsemanship; and the coaches of fine ladies, who go thither to shew their equipages" (60–61). Jerry Melford mentions a rough and uncomfortable stretch of road between Newark and Wetherby owing to "the heavy rains in the spring" (151). Then there are the accidents: the overturning of the coach between Bath and Marlborough on the way to London, which leads to the hiring of the title character's services, or the damage caused to the suspension of the coach by crossing a "deep gutter made by a torrent" on the way to Scotland (173). Such vignettes complete our understanding of what travelling by road was like in the eighteenth century. Combining as it does the cartographic outline of the route with the narrative of what happens on the road, the epistolary form of *Humphry Clinker* presents the most original and skilful treatment of geographical space in the travel fiction of the time. The novel thus demonstrates that the cartographic representation of geographical space had become a well-established norm in realistic fiction.

The depiction of the travellers' transitory occupation of the space of the road on the other hand no longer has the broad social interest that it had in Defoe's and Fielding's novels. Nor are its travellers of the same making. The social focus of *Humphry Clinker* is comparatively narrow. Except for servants, the Bramble party meet mostly people of their class and with the same way of life: members of the gentry and the upper middle-class. Their carriage and the lodgings they take isolate them from the lower classes on the road and at stopovers, confirming the fact that social segregation could be as rigorous on the road as in London. For this reason, Smollett's depiction of the road as contact zone loses what had been one of the fascinating traits of its predecessors: their depiction of the road as a space used willy-nilly by all classes. This loss is, however, compensated by the breadth of the correspondents' discussions of the development of the country and of issues such as

England's union with Scotland, religion, agriculture or education. 'At the same time, with its greater emphasis on nature, the countryside and the Highlands, the novel announces a new development in the representation of geographical space.

Matthew Bramble and his party's journey is neither motivated by social necessity nor by economic reasons: another departure from the pattern in other novels. They tour Great Britain for their pleasure: the pleasure of revisiting or discovering people and places. They are tourists in the modern sense of the term. Their journey reminds one of Moll Flanders' two jaunts to Oxford and Gloucester. Each member of the party is expected to derive some personal benefit from the tour and come back home enriched by the experience. It is in search of better health that Matthew Bramble sets out from Bramble Hall to visit the spas on his way, with little success, however. He eventually finds what he is looking for not at a spa but in Scotland. He is so delighted with its unspoilt nature that he forgets his pains and is well on his way to finding the "felicity" he has mistakenly been looking for in spas (294). For his nephew and niece, the tour is educational: the young people learn much about the country, their class, social life and themselves. Tabitha, Matthew Bramble's sister, makes use of the opportunity of the tour to unearth a husband, one more instance of the sexualisation of the road, but in a different, oldmaidishly comic key. A popular saying has it that, for most travellers, the return home is one of the highlights of a journey. This is as much the case in *Humphry Clinker* as in *Joseph Andrews*, *Tom Jones* or *Roderick Random*. Home is once again an important place in the geography of the novel. During the second part of the journey, Matthew Bramble increasingly realizes that there is no place like home and he looks forward to his return to Brambleton Hall. His account of the return journey in his letters reads like a preparation for the new life he promises to lead when he gets back home. His niece and sister both bring back home the desired husband. As for Jerry Melford, Bramble's nephew, he has shaken off the overconfidence of the university student, made new friends and acquired some social wisdom. The geographical setting, towns, cities, spas, estates, the sea and a lake play an essential role in the characters' experiences. On the other hand, the representation of their itinerary is still essentially two-dimensional, exhibiting the same cartographic character as the novels of Defoe and Fielding earlier in the century.

The multiplication of road journeys in the fiction of the eighteenth century and the representation of roads as a common space of motion and transit show that roads had become the site of human mobility *par excellence* on land. Of course, there had always been people who had travelled within their countries, and even travelled far and wide by land, but pilgrims, students, members of guilds and clergies, mercenaries, pedlars and, above all, merchants had not had the facilities at their disposal that the eighteenth century offered. The combination of an improved network of roads and organized transports encouraged movements. In the same way as opening up paths gave Crusoe a clear idea of the shape and superficies of his island, people's awareness of the road-system beyond the limits of their individual parishes gave them a better understanding of the extent of the national space.[34] When novelists portrayed their varied assortment of traditional and new travellers following the same routes while pursuing their many different goals, they bore witness to the birth of the age of modern roads and road travels, while giving new life to the art of fiction.

With their coaches, coachmen, inns, innkeepers, pedestrians and highwaymen, Defoe's, Fielding's and more generally the eighteenth-century novelists' narratives of their characters' travels have often been considered a quaintly outdated feature of their fictions, but the reader today cannot miss the modern touch in the emphasis they put on the distances many of the travellers cover, on means of transport—in particular on regular public transport that extended the travelling range of new categories of travellers among the middle and lower classes—on the greater freedom of movement that they made possible and on the road as a space of anonymity, leisure and petty criminality. The geography of the novel sharpens not only our sense of the similarities and differences between the world of Defoe's and Fielding's characters and ours but also our sense of what developments the age portended.

More generally, as pointed out in the first chapter, maps and atlases propose a broad vision of space and command over territories. With its indebtedness to the geography of roads and road journeys, the novel

34 The suggestion here is that this awareness of the national space predates the development of turnpike roads in the second part of the eighteenth century by as much as fifty years and that it is related to the classification of roads and their mapping at the end of the seventeenth century.

became a popular expression of greater practical and virtual control over the national territory and, by extension, over continental Europe as Defoe's *Roxana* or Sterne's *Sentimental Journey* remind us. Greater mobility, the occasion for contacts and the availability of affordable road books and maps regularly brought up to date meant that, for an increasing number of people from the middle and lower classes living away from urban centres like London and Bristol, life was no longer confined to one's birthplace and immediate surroundings. Local frontiers evaporated. A better acquaintance with road geography helped correct such misapprehensions as Tabitha Bramble's belief in Smollett's *Humphry Clinker* that Scotland could not be reached from England but by sea (199). Of course, this geography was no longer one of exploration and discoveries, but road travel remained adventurous. Movements were facilitated but travelling was still far from being always easy and pleasurable. It was never comfortable. All the novelists agree that roads, their coaching inns and the "superficies" of their vicinity[35] were sites of strange encounters, some insecurity and frequent accidents owing to the sorry state of even the most frequented highways, not to mention the quality of the means of transport. Yet, for fictional characters, as well as for contemporary readers of maps and atlases, the British Isles and the Continent had become better known places and places to know better.

35 Ogilby, "Preface", *Britannia*, [1].

4. The Emergence of the Rambler in English Literature

The second half of the eighteenth century witnessed a complete change of scene, of scale in the representation of geographical space and the appearance of an entirely new way of living space. The representation of smaller local spaces called for a revival of interest in the chorographic prospect in a modified, topographically more exact form often referred to as landscape. Today we would speak of views. In literature, this coming back into favour of the prospect was connected with the emergence of one more literary figure with a special relationship to geographical space, the "promeneur". Commonly associated with the change of sensibility and the related new feeling for nature, going for walks was, of course, neither a new activity nor a particularly remarkable expression of the by then well-established contemporary taste of the age for mobility. What happened was that it became the leisure activity of a certain intellectual elite, who were acutely conscious that, before being a landscape, a meadow, a wood, a river, a trout in the river or flowers on its banks, nature occupies a space of its own and that this space, like other geographical spaces, forms a part of the physical base of human existence. On the literary scene, the phenomenon was European, and some of the most famous early portrayals of the "promeneur" were not English. One was Rousseau's portrait of himself in sections of his *Confessions* (1765–70) and, above all, in *Rêveries d'un promeneur solitaire* (1776–78). Here he is, in the "Fifth Walk" of the latter work, going about his favourite activity on the Île Saint-Pierre on the Lake of Bienne in Switzerland:

> When the lake was too rough for boating, I would spend my afternoon wandering over the Island searching right and left for plants, sometimes sitting down in the most cheerful and solitary nooks to dream at my ease and sometimes on terraces and knolls to let my eyes wander over the superb and breathtaking view of the lake and its shores, crowned on one side by the nearby mountains and on the other spread out onto rich and fertile plains over which my sight extended all the way up to the more distant, bluish mountains which bordered it. [1]

1 Jean-Jacques Rousseau, *The Reveries of the Solitary Walker, Botanical Writings, and Letter to Franquières*, trans. Charles E. Butterworth (Hanover and London: Dartmouth College, 2000), 44–45.

In Germany, "der Spaziergänger", a person enjoying walks in the countryside, was celebrated by the contemporary Sturm und Drang Movement. Goethe himself devotes several poems to this leisure-time activity: "Osterspaziergang", "Im Vorübergehn" and "Gefunden", the last of which characteristically reads:

> Once in the forest
> I strolled content
> To look for nothing
> My sole intent.
>
> I saw a flower,
> Shaded and shy,
> Shining like starlight,
> Bright as an eye.
>
> I went to pluck it
> …
> And to my garden
> Carried it home,
>
> There to replant it
> Where no wind blows.
> ("Found", trans. Michael Hamburger).[2]

> Quand le lac agité ne me permettait pas la navigation, je passais mon après-midi à parcourir l'île, en herborisant à droite et à gauche; m'asseyant tantôt dans les réduits les plus riants et les plus solitaires pour y rêver à mon aise, tantôt sur les terrasses et les tertres, pour parcourir des yeux le superbe et ravissant coup d'œil du lac et de ses rivages, couronnés d'un côté par des montagnes prochaines, et de l'autre élargis en riches et fertiles plaines, dans lesquelles la vue s'étendait jusqu'aux montagnes bleuâtres plus éloignées qui la bornaient.
> (*Rêveries du promeneur solitaire*, Lausanne: Guilde du Livre, 1951), 150–51.

2 *Goethe, Selected Poems*, ed. Christopher Middleton, *Goethe: Collected Works*, vol. I (Princeton: Princeton University Press, 1994), 191, 193.

> GEFUNDEN
> Ich ging im Walde
> So für mich hin,
> Und nichts zu suchen,
> Das was mein Sinn.
>
> Im Schatten sah ich
> Ein Blümlein stehn
> Wie Sterne leuchtend
> Wie Äuglein schön.

The solitary walker is as present in English as in French or German literature, even though English has no completely satisfactory equivalent for the word "promeneur". The term "rambler" has been preferred to "stroller" to denote a person similarly enjoying walks because of its association with the countryside and its suggestion of purposefulness, even though, in the present case, the word must be understood to exclude as irrelevant the modern connotation of doing so "as part of an organized group" (OALD). In the British Isles, the rambler is usually a poet or the subject of poetry rather than connected with prose as in the case of Rousseau. Thinking of ramblers, one name that immediately comes to mind is that of William Wordsworth, himself a compulsive walker, most of whose poems have some connection or other with this activity.[3] But he was neither the only nor the first poet of the age to celebrate walking in his poetry. Samuel Taylor Coleridge, Robert Southey and Walter Scott also refer to rambles,[4] and, before them, Oliver Goldsmith had done so in *The Deserted Village* (1770) and William Cowper in *The Task* (1785). Other Pre-Romantic poets, Thomas Gray, James Beattie and Robert Burns, may not refer so explicitly to strolls, but the subject matter of their lyrics are often observations made during walks. As there is no solution of continuity between the Pre-Romantics' and the Romantics' evocations of the rambler and of the space of his walks, it has seemed preferable to consider their representations of the subject in the same chapter.[5]

Ich wollt es brechen
…
Zum Garten trug ichs
Am hübschen Haus.

Und planzt es wieder
Am stillen Ort …

Johann Wolfgang Goethe, *Gesammelte Werke*, ed. Bernt von Heiseler (Bielefeld: Sigbert Mohn, 1959), vol. I: 374–375.

3 Florence Gaillet-de Chezelles, *Wordsworth et la marche: Parcours poétique et esthétique* (Paris: Ellug, 2008).

4 For Coleridge, see below. Among Robert Southey's poems touching on the subject can be mentioned "Youth and Age", "Elegy on a Quid of Tobacco", "Written on a Sunday Morning", "To Contemplation" and "The Soldier's Wife" about a poor wanderer. Walter Scott's "Hellvellyn" can be added to the list.

5 In what follows and in the next chapter, references to the lines or stanzas of the poems discussed are given in the text and page numbers in footnotes, with the titles of the editions used.

Under the influence of voyagers' reports, maps and atlases, travel fiction had emphasized spatiality and minimized the importance of other features of the topography. Its cartographic representations were well suited to the description of vast spaces such as seas, continents, countries or even regions and to narratives dealing with travels, departures, routes, destinations and the lived experience of distances. That the space of the rambler differs in essential ways from the traveller's is obvious. It is comparatively narrow, its horizons are limited and there is the unevenness of the ground to take into consideration. As for the rambler himself, he behaves differently from the traveller in several ways. He does not know the traveller's preoccupation with reaching a destination. His main purpose is to enjoy his surroundings, and this explains why his accounts of his walks opened a new chapter in the representation of space. The change of perspective necessitated a corresponding change in the means of representation. A problem for the writer was that no satisfactory geographical model was available. Since the time of the descriptions of Crusoe's island or of Joseph Andrew's countryside, there had been little progress in the verbal representation of small-scale local topography. Two-dimensional mapping could still be suitable for drafting the plan of a town but it was inadequate when it came to representing movements among hills, mountains and valleys or through woods, fields and meadows. For Rousseau's contemporary, Bernardin de Saint-Pierre another difficulty was to find the right words to depict the character of a landscape:

> The art of describing natural objects is so little known or used, that terms to express them are not yet invented. Attempt but the description of a mountain in such a manner that it shall be recognized; when you have spoken of the base, the flanks, and the summit, you have said every thing. But what a variety is discernable in these forms, round, long, flat, hollow—'tis a mass of words without information. The difficulty is the same as to hills and vallies [sic]—[6]

6 Bernardin de Saint-Pierre, *Voyage to the Island of Mauritius, (Or Isle of France), the Isle of Bourbon, the Cape of Good-Hope*, trans. John Parish (London: 1775), 264.
L'art de rendre la nature est si nouveau que les termes mêmes ne sont pas inventés. Essayez de faire la description d'une montagne, de manière à la faire reconnaître: quand vous aurez parlé de la base, des flancs et du sommet, vous aurez tout dit. Mais que de variété dans ces formes bombées, arrondies, alongées, applaties, cavées, etc.! Vous ne trouvez que des périphrases: c'est la même difficulté pour les plaines et les

Cartographic representations being inadequate, it was possible to have recourse to the other method of geographical representation, the prospect and its derivatives, which froze-framed single moments of perception and were static. But then, what was relatively easy for a representation in a visual medium was much less so in the verbal one, as the several hundreds of topographical poems composed during the century demonstrate. To obviate this drawback, one solution suggested itself: to use geographical comments similar to those accompanying maps in guidebooks. By combining fleeting visual impressions with, or inserting them in, a sequential listing of topographical and other natural features such as they presented themselves along the way, it was possible to give graphic accounts of walks in the countryside.

The previous chapters may have given the impression that geographers themselves had abandoned prospects. In fact, if they had suffered an eclipse in travel literature, prospects had never ceased to be utilized as topographical representations. They had evolved to respond to other requirements and new tastes. Professional topographers had continued to produce detailed visual profiles of areas in the form of drawings and watercolours, their aim being still to produce accurate perspectives of cities, estates and, as noted in the introduction, possible theatres of operations for the armies. Gifted professional topographers like Edward Dayes, William Pars, Paul Sandby, Thomas Hearne and many others added artistic sensitiveness to the topographical exactness of their predecessors, topographers like Wenceslas Hollar or Nathaniel and Samuel Buck (fig. 4.1). During the second half of the eighteenth century, because it became an art, chorographic representation met with the taste of the time for landscape painting. British travellers had started bringing back to England, mainly from Italy, not only paintings by old masters, but also "views, more or less embellished, of the antiquities" and of other striking places visited.[7]

---

vallons. (*Voyage à l'Ile de France, à l'Isle de Bourbon, au Cap de Bonne Espérance, etc. par un Officier du Roi*, Seconde Partie (Neuchâtel: 1773), 151.
Bernardin de Saint-Pierre's description of Mauritius, the setting of his pastoral romance, *Paul et Virginie*, attempts to solve the problem along the same lines as the literature of the "*promeneur*" described in this chapter. On the subject, see also Paul Claval, *Histoire de la géographie*, 38.

7 Graham Reynolds, *Watercolours: A Concise History* (London: Thames and Hudson, [1971] 1998), 28–29. See also Andrew Wilton and Anne Lyles, *The Great Age of British Watercolours*, 1750–1880 (Münich: Prestel, 1993).

Fig. 4.1: Paul Sandby, "A View of Vintners at Boxley, Kent, 1794".

Fig. 4.2: John Robert Cozens, "Lake Nemi, Campagna, Italy".

Art historians usually speak of a change of taste and of a corresponding shift in painting from the tradition of Claude Lorrain's and Salvator Rosa's composite landscapes to the more literal depictions of topography of Dutch landscape painting.[8] Leading in the same direction, there was also the influence of Canaletto, whose topographical pictures of Venice became so popular with English gentlemen on the Grand Tour that they attracted him to London, where he sojourned between 1746 and 1755 to be closer to his market. In their turn, English artists went to Italy and followed his example of painting landscapes outdoors: "from nature" was the expression. It is not exaggerated to say that it was the combined impact of his work and that of native artist topographers that led English landscape painting to develop a precise geographical character. Their aim may have been to transcend what Heinrich Fuseli called the mere "view".[9] The fact nonetheless remains that their landscapes tended more and more to present an identifiable topographical delineation. The work of John Robert Cozens reconciles the two trends, the Claudian tradition and the topographical one (fig. 4.2).[10] Considered in this context, the return to favour of a similar type of prospects in poetry can be said to have kept up the connection with geography. As will be seen below, the new poetry does in places exhibit a visual quality that recalls that of contemporary landscape painting and watercolours or the vistas for which English gardens were becoming famous. It above all makes precise

8 In "Ut pictura poesis ut pictura hortus, and the picturesque" (*Word and Image*, vol. I (1985), 85–107), John Dixon Hunt identifies the same change in landscape gardening and proposes five explanations: (1) a taste for less intellectual and learned pictures, (2) to appeal to classes less concerned with antiquity, mythology and historical figures, (3) a more sentimental attitude, (4) linked to the public's and the artists' discovery of the so far neglected beauties and cultural treasures of the British Isles, and (5) a growing interest in nature itself rather than in human nature that still predominated in the Claudian tradition in painting defended by Sir Joshua Reynolds. On this last point, see Ronald Paulson, *Literary Landscape: Turner and Constable* (London: Yale University Press, 1982), 47–60.

9 Johann Heinrich Fuseli, *Lectures on Painting* (London, [1801] 1820), 179.

10 His father and mentor, Alexander Cozens was heir to the seventeenth-century Dutch and French tradition of landscape painting but, as drawing master, he also taught boys and private pupils the rudiments of topographical sketching, in particular to boys who would become naval men and officers. See Kim Sloan, *Alexander and John Robert Cozens. The Poetry of Landscape* (London: Yale University Press, 1986), 22–23, 40–50.

references to actual sites. On the other hand, it is much more diverse. Representations of landscapes there are, yet the focus of poems is just as often natural features or people that the viewer passes on his way but would not be shown on a map or would not have their place in a Claudian mythological or historical landscape.

Considering that the rambler appeared late on the eighteenth-century scene, much later than the voyager or the traveller, one may well ask whether there was a filiation between him and his predecessors and between their representations of space. Three works and three dates suggest that there was not. Laurence Sterne's *Sentimental Journey* appeared in 1768. Geographically speaking, its representation of space was still the same as that in *Joseph Andrews*: the novel told the story of Yorick's journey to Calais, Amiens, Paris, and through the Bourbonnais and Savoy. Each amorous encounter or encounter with poverty and the little miseries of life is systematically associated with the name of a place, usually a town as in *Moll Flanders*. Only twice is the word nature used to refer to the biophysical world. In the other instances the word denotes the human character.[11] In this the novel remains faithful to the classical and Augustan tradition, as well as that of travel fiction. Goldsmith's poem, *The Deserted Village*, published two years later, was the first work to describe a rambler's walk in English literature.[12] It only had an indirect connection with travel literature: the poem was supposed to be a traveller's or exile's account of his return to the village where he grew up, a village claimed back by wild nature during his long absence. The following year, in 1771, Smollett's novel *The Expedition of Humphry Clinker* discussed in the previous chapter as a cartographic journey similar to *The Sentimental Journey*, contains clear hints, but only hints, that the main characters develop a taste for strolls at the climax of their tour, during their stay on the shores of Loch Lomond in Scotland. For them and for Smollett's readers,[13] the visit to the area was literally a discovery of unspoilt nature. However, this discovery was the

11 Laurence Sterne, *A Sentimental Journey, with the Journal to Eliza and A Political Romance*, ed. Ian Jack (Oxford: Oxford University Press, [1968] 1984), 113, 120, as opposed to 49–50, 74, 84, 100, 107, 109, 112, 119, 124.

12 Although some passages of James Beattie's nearly contemporary *The Minstrel* echo a rambler's descriptions of his walks, the poem presented as a legend and allegory is, of course, not rooted in actual topography.

13 Jeremy Lewis, *Tobias Smollett* (London: Jonathan Cape, 2003), 250 & passim.

unexpected outcome of the main character's essentially practical search for health, not yet the kind of spontaneous return to nature that one has come to associate with the Romantic Movement and the cult of the Sublime.[14] As for the representation of the topography in the novel, it remains essentially flat and cartographic. The numerous references to topographical features, like rivers, are of places one passes on the way as in the traveller's geography. The author's description of mountains is reminiscent of maps. To a Matthew Bramble more interested in productive gardening and convenient architecture than in landscapes, the Highlands appear, from a distance, "heaped upon the backs of one another", a descriptive phrase that recalls the molehills indifferently symbolizing hilly and mountainous massifs.[15] On the whole, the geography of the novel, like its narrative, is the traditional one found in fiction of road journeys. The publication within four years of these three works makes it clear that one can neither speak of a progressive evolution from traveller to rambler nor of a shift of the traveller's attention to his immediate surroundings. The rambler was a new actor on the cultural scene of the eighteenth-century. He opened what I shall not hesitate to call a new space. The three works furthermore suggest that travellers and ramblers were from then on going to coexist and occasionally exchange roles, moved as they were by a similar urge to go and visit places.

The absence of filiation between travel accounts and evocations of rambles does not mean that resemblances do not exist between them. These concern the general level of representation: the use of precise geographical information and the literal throwing into relief of the rambler's movements in his space. Neither had played a role fifty years before in James Thomson's *The Seasons* (1726–1730), usually considered one of the earliest expressions of the new feeling for nature at the dawning of the Age of Sentiment. On the other hand, both traits can be noted in the passage from Rousseau's *Rêveries* quoted at the beginning of this chapter. This means that the rambler's participation in the reappraisal of nature was a late one. When he appeared on the literary scene, the return to nature was well under way. What he contributed to it was the introduction into the text of a dynamic perception of space,

14 Robert Giddings, *The Traditional Smollett* (London: Methuen, 1967), 145.
15 *The Expedition of Humphry Clinker*, 234.

similar to that afforded by the changes of scenes in travel literature. It was this addition that transformed the static, contemplative, "georgic" character of Thomson's poems, in which, as in paintings, as John Barrell shrewdly observes, only the eye travels.[16]

The utilization of geographical material was not new in poetry. Geography had even been a source of inspiration for the topographical poem, the minor poetic form referred to above and at the beginning of Chapter 2 for its connection with chorography, the scenographic art of geographical representation. That a link exists with the rambler's poetry examined in this chapter and the next is undeniable. The rambler's poetry can be seen as an attempt to breathe new life into a prolific and popular tradition that was running to seed through insipid repetitions. It reacted against its hackneyed themes, slavish imitation of models and excessive use of moralizing reflections. Above all, it went against the far too generalized character of its topographical descriptions of places.[17] Of this imprecise depiction, John Dyer's "Grongar Hill" (1726), one of the most often praised topographical poems, is a typical example. The poem presents a landscape intended to create a picturesque impression. Owing to its use of general plurals before countable nouns and of the definite article as deictic or to designate a class of objects, the author gives his description a generalized turn that blurs the geographical character and makes it look like a Salvatorial composite landscape:

> Now, I gain the mountain's brow,
> What a landskip lies below!
> …
> Old castles on the cliffs arise,
> Proudly towering in the skies!
> …
> And see the rivers how they run
> Thro' woods and meads, in shade and sun!
> …

16 John Barrell, *The Idea of Landscape and the Sense Place, 1730–1840. An Approach to the Poetry of John Clare* (Cambridge: Cambridge University Press, 1972), 10, 21–23.

17 Robert Arnold Aubin, *Topographical Poetry in XVIII-Century England* (New York: MLA, 1936), 219. The author deplores the lack of concreteness of the descriptions, what he calls the Augustan "prejudice against minute and locally characteristic portrayal" (210).

The fountain's fall, the river's flow,
The woody valleys, warm and low:
The windy summit, wild and high,
Roughly rushing in the sky![18]

An examination of the whole picture further reveals that the main topographical features are juxtaposed as on a screen, which is not a complete surprise coming as it does from a writer who was also a painter.

That in "The Sofa", "The Winter Morning Walk" and "The Winter Walk at Noon" (Books I, V and VI of *The Task*) William Cowper taps the same source of inspiration as topographical poems is evident. His poems include depictions of landscapes, fragments of landscapes and long reflections in the same proportions. The first depiction of a landscape in "The Sofa" characteristically follows the traditional descriptive procedure. It presents a succession of horizontal planes from the closest in the foreground to the horizon in the middle of the picture. The view is from above: from a higher ground—Cowper calls it an "eminence"— in what appears to be a rather flat area. Like Thomson's views, it lays emphasis, not on the mapping of the surroundings, but on the depth of the field of vision, on biophysical nature and rural activities, on the relief of the area, its river, meadows, trees and hedges, with villages lying discreetly in the background:

Thou know'st my praise of nature most sincere,
And that my raptures are not conjured up
To serve occasions of poetic pomp,
But genuine, and art partner of them all.
How oft upon yon eminence,(*) our pace
Has slacken'd to a pause, and we have borne
The ruffling wind scarce conscious that it blew,
While admiration feeding at the eye,
And still unsated, dwelt upon the scene!
Thence with what pleasure have we just discern'd
The distant plough slow-moving, and beside
His lab'ring team, that swerved not from the track,
The sturdy swain diminish'd to a boy!
Here Ouse, slow winding through a level plain
Of spacious meads with cattle sprinkled o'er,

18 John Dyer, *Grongar Hill*, ed. Richard C. Boys (Baltimore: Johns Hopkins Press, 1941), lines 41–43, 49–50, 93–94, 105–108.

Conducts the eye along the sinuous course
Delighted. There, fast rooted in his bank
Stand, never overlook'd, our fav'rite elms
That screen the herdsman's solitary hut;
While far beyond and overthwart the stream
That, as with molten glass, inlays the vale,
The sloping land recedes into the clouds;
Displaying on its varied side the grace
Of hedge-row beauties numberless, square tow'r,
Tall spire, from which the sound of cheerful bells
Just undulates upon the list'ning ear;
Groves, heaths, and smoking villages remote. (Lines 150–76)[19]

What one misses here is Thomson's Claudian subtlety, but the reader is likely to pay only glancing attention to this as, in Cowper's poems, it is no longer the eye that controls the organization but the narrator's account of a walk around the countryside. The consequences are threefold: the narrator's description of his itinerary gives new importance to the geography of the area and the identification and localization of the places; it changes the narrator's relation to the landscape; and his movements modify his perspective on the landscape at every step. The narrator's allusions to his effort to reach the top of the hillock (line 6) and to the wind blowing (line 7) in the passage quoted, as well as later references to sounds and smells, imply that the narrator is within the landscape and no longer an external observer as in paintings or in Thomson's *The Seasons*. Because the long introduction furthermore identifies the narrator with the author, readers were quick to realize that the description exactly fitted a landscape in the neighbourhood of Olney, in Buckinghamshire, where the poet lived. How geographically precise the description of the view and of the rest of the walk was is demonstrated by the footnotes in the nineteenth-century edition used here. The first footnote concerns the landscape seen from the said eminence:

(*) "From the town of Olney, westward, over three fields, the ascent is gradual to the eminence referred to by the poet in these lines. From this elevation is seen a prospect extensive in every direction but the north, which is bounded by a quick hedge, on rising ground. To the eastward is Steventon, in Bedfordshire; further east stands the 'square tower' of Clifton Church, and ranging still eastward, the prospect is bounded

19 *The Poetical Works of William Cowper*, ed. Robert Aris Willmott (London: George Routledge, 1855), 191–92.

by Clifton Wood; till, due east, is seen the 'tall spire' of Olney Church, and a considerable part of the town. To the southward is the pleasant village of Emberton, on the right of which appears, when the weather is clear, Bowbrick-hill, the church on its summit, at the distance of nearly fourteen miles. Due south, in an extensive valley, appear the devious windings of the river Ouse, whose mazy and deceptive course assumes the semblance of various streams. The meadows are likewise intersected by dykes, cut for the purpose of draining floods, which give the land, even in times of drought, a delightful verdure." —*Cowper Illustrated*, 1803, p. 31.[20]

The geographical information supplied by the editor underlines the fact that Cowper's verse description does not propose a composite or all-purpose landscape like most topographical poems but an actual view from a particular spot: a view such as artist topographers drew. He duly localizes geographical features with respect to one another—a difficult thing to do with words— and exhibits a better sense of spatiality than his predecessors. His vision is truly topographical. The rest of the narrative of the walk confirms this.

It is obvious that the two descriptions of the same landscape are not accurate in the same way. The prose description is explicitly quoted from a book that includes, beside a short biography of the poet (pages 3–26), a guide to the poet's favourite spots in the neighbourhood of Olney, the very ones the poet mentions in the account of his walk (pages 27–53). This second part of the volume is entitled "Description of Weston Park, etc." Its author, E. W. Brayley, was an antiquary and a topographer, in other words, a geographer, as well as an admirer of the poet. His *Cowper Illustrated* offers exhaustive descriptions, a systematic listing and locating of places, with greater emphasis on their naming. The volume includes pictures of the principal sights on the tour (fig. 4.3). Because their author is a geographer, there is something final about his mapping out of the area. The poet's depiction lacks this definite character. It is accurate in as much as it cannot be faulted and allows the reader to follow the rambler's itinerary. The same will be true in Romantic poems whose titles name the places that have inspired them. The landmarks, specific trees, a crossing of paths, a small bridge, a hedge, or an isolated house encountered by Cowper during the stroll, form a looser sequence than in the geographer's description.

20 E. W. Brayley, J. S. Storer and J. Creig, *Cowper, Illustrated by a Series of Views: In, or Near, the Park of Weston-Underwood, Bucks* (London, 1803). E. W. Brayley is the author of the texts and possibly the pictures; Storer and Creig are the engravers.

Fig. 4.3: E. W. Brayley, "View of Olney Church from a nearby hill" in *Cowper Illustrated.*

This may be intended to suggest that the rambler does not always follow the same paths ("Once went I forth, and found, till then unknown, / A cottage", lines 193–94). It conveys a sense of freedom of movement that the geographer's more systematic procedure kills. This less geographical and more poetic manner is natural and in keeping with the fact that the narrator, being in the landscape, records his impressions, whereas the geographer's detached attitude fixes them as facts.

It should be pointed out in passing that the descriptions of places and the engravings of the poet's favourite spots and landscapes in *Cowper Illustrated* made it the perfect rambler's guide for whoever wanted to follow in the poet's footsteps. Published eighteen years after the poem, it even mentions what had changed since Cowper's time. As the work of a topographer, it was also a convincing sample of the kind of discourse that geographers had begun, concurrently with the rambler's evocations of space, to elaborate to describe limited local settings. If ramblers' guides were still rare, another rather similar type of guide-books was offered to the public at the same period. They described more ambitious tours, with maps on the model of those of road books. These and topological surveys showed that the geographical discourse on a scale from the parish to the region was catching up with literature and the visual arts.[21] Like them, they took the relief of the ground more fully into account, announcing developments that will be considered in the next chapter.

For us today, the early readers' interest in the geographical dimension of Cowper's and other writers' verse is somewhat surprising. The publication of *Cowper Illustrated* and the use made of it, fifty years later, in the footnotes of an edition of Cowper's poems suggest that they expected and enjoyed this geographical accuracy. The recognition of well-known places seems to have formed part of their aesthetic

21 See Joseph Craddock, *Letters from Snowdon: Descriptive of a Tour through the Northern Counties of Wales* (London, 1770); Richard Warner, *A Walk through Wales in August 1797* (Bath, 1798); Richard Warner, *A Walk through the Western Counties of England* (Bath, 1800). To these could be added surveys focusing on topography: William Tunnicliff, *A Topographical Survey of the Counties of Hants, Wilts, Dorset Somerset, Devon, and Cornwall* (Salisbury, 1791) and Archibald Robertson, *A Topographical Survey of the Great Road from London to Bath and Bristol* (London, 1792). The first editions of the famous Baedeker appeared at the beginning of the nineteenth century.

experience in reading poetry or poetical prose. In this sense, literature, like painting and watercolours, responded to a general expectation of the precise localization of places. With or without the help of guidebooks like E. W. Brayley's, readers *would* try to follow the itineraries of their favourite writers. For the same reason, Goldsmith's deserted village, "Sweet Auburn, loveliest village of the plain"[22] was long believed to be Lissoy, a village in Ireland where the author spent much of his childhood. At the same time, readers could hardly have missed the fact that there were other Auburns, closer to their homes, English villages suffering a similar fate. This double reading shows that while readers expected the description of a geographical space and its landmarks to vouch for the existence of the place, they were simultaneously aware that each description had an emblematic character and was a source of symbolic meaning. This need to anchor an aesthetic experience in a concrete geographical space, be it on the scale of an estate, parish or district, is, in every respect, the same that had made the success of travel fiction on the world scale. The readers' expectation of a geographical fundament to the poetic inspiration, which encouraged their crosschecking with the life of the author, was naturally related to the curiosity about the lives of writers, which was to give birth to detailed biographical criticism in the nineteenth century.

To come back to the account of the narrator's walk in "The Sofa", more novel than its utilization of precise geographical information was the role played by the rambler's changes of perspective as he goes from spot to spot. In the above quotation, the rambler saunters up a hillock. When he reaches the top, he stops to look at the view and then moves on. This simple narrative procedure illustrates the way accounts of a rambler's movements could contribute to animating the static landscapes of the "georgic" tradition. Its main purpose was to highlight the changes of perspective on the scenery when the path changes direction or the rambler walks up or down a slope. This kinetic way of apprehending one's surroundings, of which topographers had a professional experience in land surveys and their preparations of military campaigns, is one more expression of the eighteenth-century on the move, and it is probably

22 Oliver Goldsmith, *Collected Works of Oliver Goldsmith*, ed. Arthur Friedman, vol. V, (Oxford: Clarendon, 1966), 286–303, 1.

this kinetic way of apprehending one's surroundings that has led historians of Romanticism to prefer the word landscape to prospect. The walk continues and the narrative mentions other features in the landscape, other hills and dales offering other glimpses of the area. Simultaneously, the description of the walker in movement brings the topography of the land into vivid relief to the extent that he also experiences it with his feet. And again, the nineteenth-century editor does not fail to provide footnotes testifying to the author's topographical accuracy. For the poet, the next place of interest after the eminence with its view is "a cottage" (1):

> 'Tis perch'd upon the green-hill top, but close
> Environ'd with a ring of branching elms
> That overhang the thatch, itself unseen
> Peeps at the vale below … (lines 221–25)
>
> *(1) It stood about half a mile from Weston House.*

Next comes,

> Not distant far, a length of colonnade
> … monument of ancient taste, (lines 252–53)

before

> Descending now (but cautious, lest too fast)
> A sudden steep, upon a rustic bridge (2)
> [the walker] pass[es] a gulf, in which the willows dip
> Their pendent boughs, stooping as if to drink. (Lines 266–69)
>
> *(2) The bridge spanned a brook, which, after winding through the Park [of John Courtney Throckmorton, Esq., of Weston Underwood], crossed the road from Olney to Northampton, at a place called Overs Brook.*

On the other bank

> Hence ankle-deep in moss and flow'ry thyme
> [the rambler] mount[s] again, and feel[s] at ev'ry step
> [His] foot half sunk in hillocks green and soft,
> Raised by the mole, the miner of the soil. (Lines 270–73)

Another hilltop is reached:

> The summit gain'd, behold the proud alcove (3)
> That crowns it! (Lines 278–79)
>
> *(3) A graceful little structure of wood.*[23]

From this second hill the view is, as expected, different: sheep grazing on a slope in the foreground, haymaking in the middle distance, then a grove here and sunken wood there, more hills and, further back, "The Ouse, dividing the well-watered land, / Now glitters in the sun, now retires" (lines 323–24).[24]

The same dynamic vision is found in other works of the same period. In Goldsmith's *The Deserted Village*, the narrator takes the reader along with him on a tour around the now uninhabited place. Of the houses, little remains. The suggestion seems to be that violence was done to the place, that it was razed to the ground in accordance with the owner's plans for his park rather than left to decay.

> Sunk are thy bowers, in shapeless ruin all,
> And the long grass o'ertops the mouldering wall … (lines 47–48)

The poet leads the reader first up to the "decent church that topt the neighbouring hill" (line 12)—note the use of the past tense—, down to the brook "choaked with sedges" (line 41), across neglected fields and "ruined grounds" (line 78). Approaching another eminence, the rambler recalls earlier walks when

> Sweet was the sound when oft at evening's close,
> Up yonder hill the village murmur rose;
> There, as I past with careless steps and slow,
> The mingling notes came softened from below;
> The swain responsive as the milk-maid sung,
> The sober herd that lowed to meet their young,
> The noisy geese that gabbled o'er the pool,
> The playful children just let loose from school … (lines 113–20)

The reader follows the narrator past decaying walls, past the spots where the forsaken vicarage (line 140) and schoolhouse (line 196) stood. The

23 *The Poetical Works of William Cowper*, 193, 194.

24 *The Poetical Works of William Cowper*, 195.

visit ends at a place where there was once a signpost (line 220) near the now silent ruin of the public house (line 221), two strong symbols of the village's past relations with the rest of the world and testimonies of its geographical, and consequently economic and political existence. Similar is the effect produced by Wordsworth's poems and other short Romantic lyrics. The bench in "Inscription for a seat by the pathway side ascending to Windy Brow" or "The Small Celandine" seen on a roadside become objects that the poet walks past and leaves behind.

> But lately, on a rough day, this Flower I passed,
> And recognized it, though an altered Form,
> Now standing forth an offering to the Blast,
> And buffeted at will by Rain and Storm. (Lines 9–12)[25]

To watch the frolics of a green linnet from under a certain beloved tree ("The Green Linnet") or to listen to the call of the cuckoo at a given spot ("To the Cuckoo")[26] become the aims of outings. What Wordsworth calls the "spots of time" that his lyrics sing are in fact moments when the narrative of the outing freezes and frames a spot that had a geographical existence and was for him localizable in the neighbourhood where he lived. To this English cinematographic metaphor, French prefers the more descriptive and dynamic "stop on a picture", "arrêt sur image":

> The May is come again:—how sweet
> To sit upon my Orchard-seat!
> And Birds and Flowers once more to greet,
> My last year's Friends together… ("The Green Linnet", lines 1–4)[27]

"Tintern Abbey" belongs to the same category. Like Cowper's *The Task* and *The Deserted Village*, it offers, mixed with reminiscences, a series of views of the valley as they meet the eyes during the walk.

The differences between the representations of the rambler's and the traveller's space largely prevail over their resemblances and these differences are not limited to the new emphasis of the latter on the third

25 *William Wordsworth: The Major Works*, ed. Stephen Gill (Oxford: Oxford University Press, 2000), 288.

26 *William Wordsworth: The Major Works*, 245.

27 *William Wordsworth: The Major Works*, 248.

dimension of topography. The scales of the two worlds have already been mentioned. Compared to the traveller's, the rambler's space has been seen to be narrow and restricted. Whether prompted by curiosity or the spirit of adventure, or thrown onto roads and overseas routes by the necessity of trade or unhappy circumstances, eighteenth-century travellers cross oceans and continents and roam the roads of their own or other countries. The rambler's walks can take anything between a few hours and half a day. The pace is leisurely. The space covered is limited. Unlike the traveller, the rambler explores his immediate surroundings. Yet the narrowness of his space does not give the impression of confinement. On the contrary, a sense of freedom of movement, often absent in travelling tales, pervades the experience: a note subtly caught by the unexpectedly light rhythm of the iambic pentameter of the first lines of *The Deserted Village* as the speaker reminisces the village of his youth:

> Sweet Auburn! lovliest village of the plain,
> Where health and plenty cheared the labouring swain,
> Where smiling spring its earliest visit paid,
> And parting summer's lingering blooms delayed:
> Dear lovely bowers of innocence and ease,
> Seats of my youth, when every sport could please,
> How often have I loitered o'er thy green … (lines 1–7)

In the rambler's space, closeness is synonymous with intimacy. Favourite spots frequently seem to be, like Goldsmith's, situated in hollows. The archetypal poem could once more be said to be Wordsworth's valley of the Wye above Tintern Abbey.[28] The following illustration, however, comes from Coleridge. Narrating an outing from a place of retirement that he calls "The Valley of Seclusion", another valley, the poet, in a passage recalling the first quotation from Cowper, describes his climb to the top of a mountain nearby:

> But the time, when first
> From that low Dell, steep up the stony Mount
> I climb'd with perilous toil and reach'd the top,
> Oh! what a goodly scene! *Here* the bleak mount,
> The bare bleak mountain speckled thin with sheep;
> Grey clouds, that shadowing spot the sunny fields;

28 *William Wordsworth: The Major Works*, 131.

And river, now with bushy rocks o'er-brow'd,
Now winding bright and full, with naked banks;
And seats, and lawns, the Abbey and the wood,
And cots, and hamlets, and faint city-spire;
The Channel *there*, the Islands and white sails,
Dim coasts, and cloud-like hills, and shoreless Ocean—
("Reflections on having left a place of retirement", lines 26–37)[29]

The panorama is wide, the horizon boundless, but the prospect lays the main emphasis on the valley, the river, the meadows and the human dwellings below the hill. The sea marks the boundary of the local setting. It is interesting to note in passing that whereas Wordsworth prefers the valleys, Coleridge, like Cowper, usually draws inspiration from climbing expeditions—a distinctive trait of character? In "Lines: Composed while climbing the left ascent of Brockley Coomb, Somersetshire, May 1795" he writes, again highlighting the depth of the field of vision:

With many a pause and oft reverted eye
I climb the Coomb's ascent …

Beneath the dark green boughs
(Mid which the May-thorn blends its blossoms white)
Where broad smooth stones jut out in mossy seats,
I rest:—and now have gain'd the topmost site.
Ah! what a luxury of landscape meets
My gaze! Proud towers, and Cots more dear to me,
Elm-shadow'd Fields, and prospect-bounding Sea! (lines 1–2, 8–14)[30]

Whether the traveller is bound for distant countries or places closer to his home, his destination, like his route, is often unfamiliar to him. He is in search of, or confronted with, new impressions. The places the rambler visits are usually familiar to him. He feels a strong attachment to them. The rambler is fond of what he sees and is a close observer of his surroundings. His attachment to them is once more nicely described by Coleridge in an early and still somewhat immature lyric entitled "A Wish: Written in Jesus Wood, Feb. 10, 1792". The poet is following the sinuous course of a stream:

29 *Coleridge: Poetical Works*, ed. E.H. Coleridge (Oxford: Oxford University Press, 1969), 107.

30 *Coleridge: Poetical Works*, 94.

Lo! through the dusky silence of the groves,
Thro' vales irriguous, and thro' green retreats,
With languid murmur creeps the placid stream
        And works its secret way.

Awhile meand'ring round its native fields
It rolls the playful wave and winds its flight:
Then downward flowing with awaken'd speed
        Embosoms in the Deep!

Thus thro' its silent tenor may my Life
Smooth its meek stream by sordid wealth unclogg'd,
Alike unconscious of forensic storms,
        And Glory's blood-stain'd palm! (Lines 1–12)[31]

The same sense of familiar reality pervades *The Deserted Village*. The visitor knows what he is going to see at every bend of the path. It is true that the place has changed, but it does not matter: the church, the parsonage, the school, the pub and the cottages may no longer exist but, as soon as he sees the site, his memory can bring them back to life:

  Near yonder copse, where once the garden smil'd,
And still where many a garden flower grows wild;
There, where a few torn shrubs the place disclose,
The village preacher's modest mansion rose.
…
 Near yonder thorn, that lifts its head on high,
Where once the signpost caught the passing eye,
Low lies that house where nut-brown draughts inspired,
Where greybeard Mirth and smiling Toil retired,
Where village statesmen talked with looks profound,
And news much older than their ale went round. (Lines 137–140, 219–224)

The area being well known and strictly delimited, the landscape is reassuring and gives an impression of cosiness that fills the heart with delight. In Goldsmith's poem, the memory of happy days momentarily prevails over nostalgia and disappointment at finding the place deserted and ruined.

Familiar are also, in the rambler's space, vegetation, animals and the details of rural activity. The writer describes them with loving precision.

31 *Coleridge: Poetical Works*, 33.

Flowers, plants, trees, and birds' songs, must be named and described like the trees that Cowper comes across in one of his walks:

> Here the gray smooth trunks
> Of ash, or lime, or beech, distinctly shine,
> Within the twilight of their distant shades;
> There, lost behind a rising ground, the wood
> Seems sunk, and shorten'd to its topmost boughs. (Lines 302–305)[32]

The rambler's space is that of an aficionado of nature. He himself belongs to a period when the vogue to botanise had become so widespread that the mention of this pastime cropped up in the most unlikely places. Is not Ann Radcliffe guilty of anachronism in *The Mysteries of Udolpho* (1794) when she makes two of her sixteenth-century gentlemen, Monsieur St. Aubert, and M. Barreaux, his neighbour, fond of it,[33] and when she scrupulously lists the trees her characters come across in their slow progress from Gascony to Languedoc at the beginning of the novel?[34] With the rambler, the reader feels far away from Defoe's or Fielding's swallowers of space and distances. The rambler is a relisher of proximate space, who looks at his surroundings as if it was a park or botanical garden.

He is attached to the space of his walks and this space is familiar to him because it is a space that he likes returning to. The journey or voyage follows a linear trajectory to a destination. Even when they are repeated like Moll Flanders' four movements between London and Chester, it is the destination and the different occasion that matter. With the rambler, each visit is a rediscovery of the same places. Compared to the traveller's, his itineraries could be said to follow a circular pattern. The reference to this repetition of the walk is current in Pre-Romantic and Romantic poetry. *The Deserted Village* is presented as a visit that the poet makes to the place of earlier walks. The account of the first walk in *The Task* specifies that the poet had been "oft upon yon eminence". Wordsworth's "The Small Celandine" refers to previous walks during which the poet had seen the flower bud open to the sun and close to the

32 *The Task*, *The Poetical Works of William Cowper*, 195.

33 *The Mysteries of Udolpho*, ed. Bonamy Dobrée (Oxford: Oxford University Press, 1989), 3, 27.

34 *The Mysteries of Udolpho*, 30, 35–36, 42, 50.

rain. "The Green Linnet" mentions a favourite spot where he liked to sit each spring. "Tintern Abbey" characteristically begins:

Five years have past; five summers, with the length
Of five long winters! and again I hear
These waters, rolling from their mountain-springs
With a sweet inland murmur. — Once again
Do I behold these steep and lofty cliffs,
Which on a wild secluded scene impress
Thoughts of more deep seclusion; and connect
The landscape with the quiet of the sky. (Lines 1–8)[35]

And, in a further passage echoing another page of Cowper's *The Task*,[36] the poet adds that this walk reminds him of many others in "boyish days" (line 73). In some lyrics, the walk may be relived only in imagination, as in Coleridge's "Reflections on having left a place of retirement" quoted above. In travel literature, time means the number of months it takes to cross an ocean or the number of days necessary to reach a destination by road. This time dimension does not weigh with the rambler, who has time to spare to enjoy his surroundings. The absence of pressure always urging the traveller forward also favours, in his case, the work of memory.

In *The Deserted Village*, the role of memory is to bring back to life the place and its inhabitants as the speaker of the poem knew them in bygone days. The walk in space becomes a walk in the past, a past happier than the present. Memory opens another time dimension. Each stop during the visit becomes a station of a pilgrimage. To use a musical analogy for what it is worth, each station at a well-remembered spot can be said to sound harmonics, and these harmonics end up ringing louder than the fundamental sad note struck by the story line of the present visit. Here the rambler is passing by what was the village school:

Beside yon straggling fence that skirts the way,
With blossomed furze unprofitably gay,
There, in his noisy mansion, skill'd to rule,
The village master taught his little school;
A man severe he was, and stern to view,
I knew him well, and every truant knew;

35 *William Wordsworth: The Major Works*, 131.

36 *The Task, The Poetical Works of William Cowper*, 190–91.

Well had the boding tremblers learned to trace
The day's disasters in his morning face;
Full well they laugh'd with counterfeited glee,
At all his jokes, for many a joke had he … (lines 193–201)

It is the repeated contrast between, on the one hand, a past which comes alive in the poet's memory, with the inhabitants of the village attending to their work, and, on the other, the emptiness and silence of the place through which he wanders that makes the poem particularly poignant. The "hollow-sounding" cry of a wild bittern and the fluttering of a bird's wings are the only audible signs of life (lines 44, 45). The place seems dead:

But now the sounds of population fail,
No chearful murmurs fluctuate in the gale,
No busy steps the grass-grown foot-way tread,
For all the bloomy flush of life is fled. (Lines 125–28)

In Cowper, the rambler's return to a place allows him to convince himself of the permanence of the rural landscape and of nature in a world of changes. From high grounds, the favourite destination of his strolls, the view remains forever the same, with the Ouse winding its course in the distance. Permanence is also a trait of the vision of life proposed by Thomas Gray's "Elegy written in a Country Churchyard" composed in a rhythm that seems to echo the melancholy step of a walker. It focuses on the unchanging pattern of life of the inhabitants of a village and on the tombs where their ancestors sleep, now oblivious to the lapse of time. Repeated walks and returns to loved places entail re-cognition: a process absolutely different from the traveller's discovery of the new. Travelling satisfies human curiosity. Iterative walks give a sense of rootedness and of participation in an unchanging pattern of life. Participation and rootedness may be somewhat artificial because essentially aesthetic yet, by awakening memories, it can become the source of a no less real mental experience.

The rambler's return to familiar places simultaneously heightens his own sense of time passing and of the gains and losses of intervening months or years. And there is no doubt that the confrontation with a certain permanence of nature adds to his awareness of the passage of time. "Tintern Abbey" contrasts the energy of the boy for whom "the tall rock, / The mountain, and the deep and gloomy wood, / Their colours and their forms, [which] were then to [him] / An appetite" (lines 77–80) with the contemplative stance of the mature man, who has learnt to love them as "The guide, the guardian of [his] heart, and soul / Of all [his] moral being." (Lines 110–11)[37] In the poem, impulsive youth has been succeeded by chastened maturity and the boy's animal enjoyment of nature by the serene contemplation of the adult. Frequently, however, a note of sadness creeps into the evocation of such visits.

> Dear native Brook! wild Streamlet of the West!
> How many various-fated years have past,
> What happy and what mournful hours, since last
> I skimm'd the smooth thin stone along thy breast ... (lines 1–4)

exclaims Coleridge in "Sonnet: To the River Otter", and the poem concludes: "Ah! that once more I were a careless Child!" (Line 14).[38] A much stronger feeling of nostalgia, bordering on bitterness, had already pervaded Goldsmith's account of his visit of the now deserted village of his youth:

> Sweet Auburn! parent of the blissful hour,
> Thy glades forlorn confess the tyrant's power.
> Here, as I take my solitary rounds,
> Amidst thy tangling walks, and ruined grounds,
> And, many a year elapsed, return to view
> Where once the cottage stood, the hawthorn grew,
> Remembrance wakes with all her busy train,
> Swells at my breast, and turns the past to pain. (Lines 75–82)

Each portrait of those who had been the life and soul of the village further deepens the sense of irreparable loss. In *The Deserted Village*, however, man, not time, is the culprit. Consoling or not, the experience of autobiographical and emotional time is a fundamental component of the rambler's experience of his space.

Awareness of time passing and the loving attention that the rambler gives his surroundings find further expression in the writers' meditations that they kindle. These however break with the tradition of reflections

37 *William Wordsworth: The Major Works*, 133, 134.

38 *Coleridge: Poetical Works*, 48.

centred on the past of the place in topographical poems. They are refocused on the mood of the walk and take a more personal at the same time as general turn. The most outstanding illustration of this aspect of the experience is, of course, Rousseau's *Rêveries d'un promeneur solitaire*. The British Pre-Romantic rambler's meditations range widely over their time: over rural activities, the humble joys and great suffering of small farmers, over victims of economic changes; over freedom, death, God, the practice of religion, philosophy; over the evils of the age, war, politics, wealth, corruption; and, of course, more simply, over the contrast between life in the country and in the city. Cowper's "Winter Walk at Noon" illustrates this process of associations of impressions. Hearing the bell of a neighbouring church, he writes,

> I again perceive
> The soothing influence of the wafted strains,
> And settle in soft musings, as I tread
> The walk still verdant under oaks and elms,
> Whose outspread branches overarch the glade. (Lines 67–71)

A few lines further down, he adds: "Meditation here / May think down hours to moments. Here the heart / May give an useful lesson to the head, / And learning wiser grow without his books." (Lines 84–87)[39] The poets' reflections are distilled by the poems as new impressions assail their senses. They tend to take a different turn in Pre-Romantic and Romantic poetry. With the former, they remain earth-bound. Goldsmith, Thomas Gray and Cowper echo the socio-economic concerns of the eighteenth century. They are critical of many aspects of the life of the time. The evocation of the past in *The Deserted Village* is a protest against the privatizing of common land that has laid waste a once happy hamlet and productive land:

> A time there was, ere England's griefs began,
> When every rood of ground maintained its man;
> For him light labour spread her wholesome store,
> Just gave what life required, but gave no more.
> His best companions, innocence and health;
> And his best riches, ignorance of wealth.

39 *The Task*, *The Poetical Works of William Cowper*, 289.

But times are altered; Trade's unfeeling train
Usurp the land and dispossess the swain;
Along the lawn, where scattered hamlets rose,
Unwieldy wealth, and cumbrous pomp repose;
And every want to oppulence allied,
And every pang that folly pays to pride. (Lines 57–68)

"Elegy written in a Country Churchyard" similarly compares the humble joys and pains of the poor in the country to those of the rich in the city to the advantage of the former, and *The Task* finds winter in the country preferable to winter in the city:

But are not wholesome airs, though unperfumed
By roses, and clear suns, though scarcely felt,
And groves, if unharmonious, yet secure
From clamour, and whose very silence charms,
To be preferr'd to smoke,—to the eclipse
That Metropolitan volcanoes make,
Whose Stygian throats breathe darkness all day long,
And to the stir of commerce, driving slow,
And thund'ring loud, with his ten thousand wheels? (Lines 732–40)[40]

This does not mean that socio-economic concerns are absent in Romantic poetry. The poor that Wordsworth meets on his walks and celebrates in such lyrics and narrative poems as "Old Man Travelling", "The Old Cumberland Beggar", or "Michael",[41] are sufficient proof that this is not the case. Yet the meditations sparked off by the poet's walks tend to be more personal. The last stanza of Coleridge's "A Wish: Written in Jesus Wood, Feb. 10,1792" quoted above is an illustration of such personal reflections. And from personal they rise to philosophical considerations. Wordsworth's "The Small Celandine" combines the observation of the plant with a moral lesson that he draws from it, applying it first to himself before applying it to the human lot. On a previous walk, the narrator of the experience had admired the capacity of the small flower to close its corolla to survive spells of rainy weather. Seeing that, half withered, it has lost this resilience to bad weather, he reflects on the change:

40 "The Garden", *The Task, The Poetical Works of William Cowper*, 243–44.
41 *William Wordsworth: The Major Works,* 29, 49–54, 224–236.

'It doth not love the shower, nor seek the cold:
This neither is its courage nor its choice,
But its necessity in being old.

The sunshine may not bless it, nor the dew;
It cannot help itself in its decay;
Stiff in its members, withered, changed of hue.'
And, in my spleen, I smiled that it was grey.

To be a Prodigal's Favourite—then, worse truth,
A Miser's Pensioner—behold our lot!
O Man! that from thy fair and shining youth
Age might but take the things Youth needed not! (Lines 14–24)[42]

Nor is all personal consideration absent from Pre-Romantic verse. Goldsmith's depiction of the destruction of a traditional pattern of life is presented as something that concerns the narrator directly as he had considered finding a peaceful retreat in "Sweet Auburn", a retreat similar to that which Cowper describes in *The Task*. *The Deserted Village* expresses the rambler's sense of personal as well as collective loss.

In all my wanderings round this world of care,
In all my griefs—and God has given my share—
I still had hopes my latest hours to crown,
Amidst these humble bowers to lay me down … (lines 83–86)

In Pre-Romantic poems, however, it is not the personal concern that is foregrounded but the censure of evils or objects of dissatisfaction with contemporary life. In so far as the rambler's movements favour an awareness of time passing and meditations, his narrow space can be said to gain in historical or social and, later, in philosophical scope what it loses in spatial extension compared to the traveller's space.

For the rambler's meditation to unfold solitude is necessary. Rambling is a solitary activity. At sea, on the road and on the continents, islands and in the towns that are his destinations, the traveller is confronted with other people and, very often, people different from him or from those he knows in the community in which he lived. Going for solitary walks allows the rambler to commune with himself and his surroundings. This does not make him a loner, however, and the solitude need not be absolute. For Cowper, walking on his own is a "seeming

42 *William Wordsworth: The Major Works*, 288.

sweet" and must be temporary to be acceptable: "My visit still, but never mine abode." (Lines 249, 251)[43] He also alludes to other walks taken with a friend. At any rate, signs of human presence are never far to seek in the space of the rambler's walks. Whether he is climbing to the top of a hill to admire the scenery or strolling along a lakeshore or a stream, he observes signs of this presence, even though they are preferably seen or heard from some distance. "The Solitary Reaper" begins:

> Behold her, single in the field,
> Yon solitary Highland Lass!
> Reaping and singing by herself;
> Stop here, or gently pass! (Lines 1–4)[44]

Occasionally, the solitary walker crosses the path of other solitary figures, a lonely shepherd or farm labourer. Wordsworth's "Resolution and Independence" describes the encounter with the famous "Leech-gatherer on the lonely moor".[45] Some of these figures are those of people marginalized by society, a vagrant or an old crippled campaigner. In *The Deserted Village*, the only inhabitant left is a silent figure glimpsed from afar, an old woman abandoned by society:

> … all the bloomy flush of life is fled.
> All but yon widowed, solitary thing
> That feebly bends beside the plashy spring;
> She, wretched matron, forced, in age, for bread,
> To strip the brook with mantling cresses spread,
> To pick her wintry faggot from the thorn … (128 – 133)

This last sign of human activity in the village is presented as a survival from the past, soon destined to disappear. It heightens the painful impression of disaster. Wordsworth can find comfort in the places he revisits, a comfort denied to the visitor in Goldsmith's poem. For the latter, the road back to the traditional rural life of yore is blocked, and the recollections afford little consolation. *The Deserted Village* is a poem about a lost paradise. The note struck shows the author at his most romantic, closer to Coleridge than to Wordsworth: the return to the beloved place is, in a sense, a last visit.

43 *The Task, The Poetical Works of William Cowper*, 194.

44 *William Wordsworth: The Major Works,* 319.

45 *William Wordsworth: The Major Works*, 260–64.

In Goldsmith's *The Deserted Village*, this trait is brought into relief by the author's handling of space, and this handling of space is of particular interest to us here because it touches upon notions at the centre of the present study. The poem in fact describes, in order to oppose them, no less than four ways of envisaging space: four types of productions of space current in the eighteenth century. It is also between two of them that the violence that destroyed the village once broke. As seen above, the basic narrative pattern, the story of the walk around the ruins of the village, defines the length and breadth of the rambler's space on the day of his visit. In this first narrow space, he is an unexpected visitor, perhaps even an undesirable one. The place is for him a source of aesthetic and emotional experience. But, as already mentioned, the narrator also admits that he had dreamt of settling in the village, thinking that it would be a suitable place to retire from the world. In his mind, this second space, the space of his imagined retirement would have been larger. It would have included the village and the surrounding countryside in a way reminiscent of William Cowper's life at Olney in *The Task*. The visitor's production of this second space, namely his intended use of it, would not have interfered with the traditional activities of the old inhabitants. He would not have made any claim on the land itself, content as he would have been to roam its lanes and paths. The third space mentioned by the poem is precisely that of the old inhabitants. It includes the village and the fields and meadows from which they drew their livelihood. It is larger than the first space and smaller than the second. As for the fourth space, that of the park which has swallowed the village and its neighbourhood, its extent is not specified but the reader easily figures out that it is considerable as it encompasses the space once farmed by the villagers. One of the central ideas of the poem, the subject of the story of what happened to the village, is that the production of some of these spaces are compatible while others are not.

During the narrator's long absence, the village has been the victim of a conflict between two of the four conceptions of the human production of space. The conflict opposed on the one hand the rural community that had from time immemorial lived on the land of the third space and on the other, the landowner who evicted the inhabitants to turn the whole area into a private park, the fourth space. The rural community's production of space had consisted in cultivating the land around the village for the benefit of its members and perhaps other communities in the neighbourhood. The rich landowner's production of space has led to the disappearance of agricultural production and the inclusion of its once agricultural space into the space of his vast park. He has also deprived the rambler of his hoped-for strolls in the region. The poem blames above all the landlord's egoism for the eradication of Auburn and inveighs against such tyrannical "sons of wealth" (line 307) who develop private parks of "extended bounds" (line 277), with even artificial lakes (line 277) in the style of the landscape gardener and architect, "Capability" Brown, for the purpose of creating "vistas [that] strike" (line 298) and can be admired from their palaces or for the purpose of affording a single family and their friends the refined, exclusive and egoistical pleasure of still another way of walking around the country: a pastime called "taking a turn" in the park: a pastime, by the way, not entirely different from the rambler's. What the poem condemns is not the strolls but a development consisting in turning a fruitful countryside into a private unproductive space, a "garden, and a grave" (line 302). This attack can be said indirectly to have dealt a belated deathblow to the idealization of the country-house and to the minor genre that celebrated it.

The central narrative of the rambler's visit to the village is moreover told in such a way that it superposes two images of this village, that of the site at the moment of the narrator's visit, which is by the same token that of a section of the new park and that of the active village of the past as his memory brings it back to life. This archaeology of the two states of the village, one brought back to life from the present ruins, is, however, not the last word on the subject. The poem alludes to the social cost of the privatisation of common land. Among other things, it suggests that it has destroyed the old social solidarity. In a picture of yet a fifth representation of space, which forms a startling contrast with the other four, Goldsmith shows how the landlord's eviction of his tenants has thrown the latter on the roads and into exile, making of them the most pitiful category of eighteenth-century travellers, namely exiles and vagrants. Their fate recalls that of Defoe's luckless characters, Moll Flanders, Captain Singleton and Colonel Jacque:

> ... To distant climes, a dreary scene,
> Where half the convex world intrudes between,
> Through torrid tracts with fainting steps they go,
> Where wild Altama murmurs to their woe.
> Far different there from all that charm'd before,
> The various terrors of that horrid shore;
> Those blazing suns that dart a downward ray,
> And fiercely shed intolerable day;
> Those matted woods where birds forget to sing,
> But silent bats in drowsy clusters cling,
> Those poisonous fields with rank luxuriance crowned,
> Where the dark scorpion gathers death around;
> Where at each step the stranger fears to wake
> The rattling terrors of the vengeful snake ... (lines 341–54)

By comparison with the harsh experience of these poor exiled peasants, the country pleasures of the landlord's family and his guests weigh very little, do they not? And the note struck by these lines jars with the general elegiac melancholy of the rambler. The criticism that the poem levels at the landowner—and also indirectly at the rambler himself who, after all, cannot completely conceal the fact that he enjoys his walk—is disturbing. Typically a Pre-Romantic feature of the poem, Goldsmith's handling of space offers a manifold reflection on the subject of man's production of space as understood by his time and a fascinating reflection on an aspect of the social climate as well as on the change in sensibility. Preceding by six years *Inquiry into the Nature and Causes of the Wealth of Nations*, which vaunts the benefits of private property, *The Deserted Village* is a scathing attack against the egoism that Adam Smith was to turn into a fashionable economic factor.

A latecomer on the scene of eighteenth-century characters on the move, the rambler possesses traits that sharply distinguish him from all the others and make him a representative of a new generation. Fond of solitary meditation and contemplative by nature, he is cousin to Henry Mackenzie's "Man of Feeling", a man of the Age of Sentiment. A lover of nature in rural England, he indulges in his favourite activity in a space where meadows, fields, woods and rivers predominate over the human habitat. Conducive as it is to his "rêveries", this nature, which bears the mark of man's activities, however fails to prevent the intrusion of the social, economic and political world on his meditation. His awareness that the space of his walks is an island of tranquillity in a sea of troubles keeps his social consciousness awake. Such is the familiar portrait of the Pre-Romantic rambler in literary history. In this respect, his relationship to the space of nature has not escaped the blame that he is an egotist preferring his privacy to any active involvement in the life of his times. Has not the rambler just been said to be, like owners of parks, a member of an educated and privileged class? That his attitude was open to criticism did not escape the contemporaries. Cowper fends off one of the reproaches that could be levelled at him when he concludes *The Task* with a vindication of his inspiration "from the charge of uselessness":

> Ask him, indeed, what trophies he has raised,
> Or what achievements of immortal fame
> He purposes, and he shall answer —None. (Lines 932–34)

The excuse he advances for his unproductive rambles is the pleasure they procure, a pleasure, he says, that lifts the soul in a way similar to religion:

> He is a happy man, whose life ev'n now
> Shows somewhat of that happier life to come … (lines 906–07)[46]

But the rambler's relationship to nature considered as a base of human activities demonstrates that things are not so simple.

His way of envisaging space and living it is in a sense no less than revolutionary compared to that of the traditional occupants of the soil, but also of voyagers, travellers, the Crusoes, Gullivers, Joseph Andrews and Matthew Brambles of preceding chapters. Unlike them, the rambler makes no social or economic claim on the space of his walks. He is not in search of a *Lebensraum* to make use of. He does not want to appropriate any parcel of it. His walks do not even impinge upon the pattern of life of the inhabitants of the rural space, a point vividly made by both Cowper and Goldsmith: to the rambler, the space of his walks is strictly a source of enjoyment. As a visitor of the places he walks about, the rambler's possession is not only immaterial but also temporary. It

46 "The Winter Walk at Noon", *The Task*, *The Poetical Works of William Cowper*, 308, 287.

leaves no trace. He breathes the air, admires what he sees and hears and, from these perceptions of his senses, he derives, beside pleasures, a feeling of peace and physical wellbeing. The ties that he develops to nature are, as pointed out in the discussion, aesthetic and emotional. Seen in this context of the production of a specific geographical space, what appears to be the expression of the often discussed egotism of many a Pre-Romantic and Romantic writer and what, to Cowper, was an activity that he had to defend against attacks of pointlessness, is in effect an assertion of the new individualism for which the age is known. It is only remotely related to the assertion of self-reliance and quest for economic autonomy celebrated in Defoe's novels. On the other hand, it presupposes this economic autonomy. The aesthetic and emotional production of space it pursues is not presented as a substitute for the production of the social space that was Swift's or Smollett's subject. Cowper's address to a rambler friend is explicit about this. The highlighting of the rambler's solitariness shows that it fulfils a need for the development, here not of social privacy—a demand for which the earlier part of the century had made room—, but of intellectual and spiritual independence for the self.

The examination of the wanderer's representations of his sphere of activity would be incomplete without a few remarks on his preference for a specific type of landscapes. Their purpose is to round off what has been said of the place of the rambler and of the representations of his space in the cultural context of the time. Novel as it is, the rambler's kinetic way of representing landscapes, with its emphasis on perspective and the relief of the land, exhibits in its application what can only be called a lack of variety. This lack of variety is all the more surprising as the spaces where nature reigned undisputed in the British Isles were vast and diverse. The spaces that the rambler prefers for his walks are, in their majority, situated in smiling rural areas, usually in the southern half of England. This is not intended as a criticism but only as a fact that cannot be ignored. The choice can be interpreted as a survival of the "georgic" tradition, a trait that would have appealed to readers beyond the circle of those sensitive to the call of nature. In this sense, the geographical precision offered some concrete compensation for the occasionally flat poetic diction of Pre-Romantic writers in particular. It led literature on the path of the type of geographical representations that announced the future tourist and hiker's guides—no mean achievement this. Last but not least, this new topographical poetry breathed the air of the natural sciences, another source of visual concreteness. The eighteenth century has been called "an age of natural history as much as an age of Newtonism."[47] For the Royal Society that promoted its study, natural history was not a simple matter of map making, classifying species and puzzling over the history of the earth in closets. The study of the earth was supposed to start with explorations on the terrain and nature received visitors other than ramblers and artist topographers. Geographers were a party to this undertaking, as were naturalists. The most popular and widespread branch of natural history at the time was botany. That botanizing became a vogue has been mentioned. Now most botanists were amateurs like Rousseau and ramblers. We have seen that, for their part, ramblers frequently manifested an interest in trees, plants and flowers.[48] It was as though the discovery of extraordinary types of vegetation on remote continents and islands by the voyagers of the seventeenth century had led the eighteenth century to the rediscovery of native vegetation.

On its small scale, the rural space of the rambler was part of the same space of nature that some great contemporary naturalists were travelling far out to discover. The focus of their writings may have been different, as the Pre-Romantic and Romantic writers were inclined to dwell on the familiar and the naturalists on what was new to them, yet their representations sprang from a common conception of what nature was and where to find it. For them, it was this biophysical base of all existence that has been mentioned several times in what precedes, the source of all life and movement, forever changing and yet perennial. They found this nature away from towns and cities. Writers stressed the opposition between its space and the space where the hand of man predominated: an opposition that was day-by-day beginning to be flagrant in some areas. Apart from sharing this common conception of nature, the Pre-Romantic and Romantic rambler and the naturalist in the field showed similar disinterested and non-utilitarian attitudes to it.

47 Philip R. Sloan, "natural history", *The Penguin Dictionary of Eighteenth-Century History*, eds Jeremy Black and Roy Porter (Harmondsworth: Penguin, [1994] 1996), 495.

48 It is no accident that Rousseau's botanical writings are often included in the same volume as *The Reveries of the Solitary Walker*.

As visitors or prospectors, they looked at nature from inside as well as from outside, their approach to it being predominantly visual. They demonstrated the same passion for its observation and depiction, so that it is not exaggerated to say that literature modestly and indirectly participated in what was becoming with scientists like Sir Joseph Banks, E. A. W. von Zimmermann and Alexander von Humboldt biogeography, the study of the way not only men but also living creatures and vegetation produce their own space and shape different landscapes.[49] At the same time, the rambler's attempt to reconcile topographical description and the naturalist's observation of nature with movement opened the door to more ambitious representations on a larger scale.

49 The case of E. A. W. von Zimmermann is particularly interesting. Geographer, biologist, philosopher as well as a great traveller, he was the author of *Geographische Geschichte des Menschen und der allgemein verbreiteten vierfüßigen Thiere* (1778–1783), which was to exert an influence on Georges Louis Leclerc de Buffon's own masterwork, his *Histoire naturelle*.

5. From the Pre-Romantic Rambler's Space to that of the Romantic Wanderer[1]

Between 1790 and 1830, a period often considered to have marked the end of the long eighteenth century, the literary representation of geographical space in literature underwent one more development. The traveller's and the rambler's geographies converged to produce representations of space that included both cartographic and chorographic features. From a retreat, nature became the object of a quest. It was as though contemporary social and political unrest, economic difficulties, a growing dissatisfaction with urban life, or what a contemporary review[2] described as an excess of civilization and refinement had awakened or exacerbated a desire for a more authentic nature than the English countryside offered with its landscapes moulded by human activities. At any rate, it led Romantic writers to roam further afield in search of this primordial nature. In the wake of voyagers, travellers and the more adventurous naturalists of their or the previous generation they found it in the wilder areas of the British Isles, the mountains of the Lake District, Wales and Scotland, or in the Alps, the Pyrenees, the Mediterranean countries and even Asia. Such explorations broadened their concept of nature and extended its superficies so as to include more extreme aspects of biophysical reality. Resuming travelling, the Pre-Romantic rambler became the Romantic wanderer and, like him, a European figure, albeit a more polymorphous one.

This more authentic nature that the Romantic wanderer went in quest of was again, basically, a geographical space. It however assumed a different aspect from its predecessor, the rambler's nature. For the Pre-Romantics, nature was the vast space of the rural countryside in contrast with the sphere of modern social, economic and mainly urban

1 The pages on the sea as a feature of nature first appeared in *1650–1850: Ideas, Æsthetics, and Inquiries in the Early Modern Era*, 16 (2009): 133–149 under the Title "From Cartographic to Topographic Representation of the Sea".

2 Review of "Essay on the Prevention and Cure of Insanity" by George Nesse Hill, *Quarterly Review*, 30 (July 1816): 398.

activities. In other words, it was one term in a fairly abstract opposition between the two entities of town and country. For the Romantics, the structure of space was more complex. It broke up into a multiplicity of spaces: spaces of social life opposed to a large variety of spaces of nature scattered around the globe. This plurality of spaces is a distinctive trait of the Romantic representation of the geographical space. Moreover, the wanderer often travelled to countries where rural life had remained "Arcadian", namely close to what he considered primordial nature.[3] As he was equally interested in the past and in things cultural, like artist topographers, it is not always easy to distinguish between his fascination with the past, the arts and his love of nature. In poetry and prose, what belonged to the past and what belonged to the spaces of nature proper coalesced when he came across antique ruins and places to which mythological reminiscences were attached. After all, Greek temples or mediaeval castles and monasteries were often located in natural settings of great beauty that appeared to have hardly changed over time. Romantic works readily associate the geography of the naturalists and cultural geography.

The English Romantic wanderer travelled eastwards, northwards and southwards in search of his nature but, interestingly enough, not westwards to Defoe's Americas: there is no equivalent in English literature to Chateaubriand's descriptions of Niagara Falls and the forests of the new Continent.[4] Indeed, British writers looked in the same direction as the promoters of England's colonial expansion in the nineteenth century. Nor is their Asia the literary Orient of *Arabian Night's Entertainments*, Beckford's *Vathek* or Goethe's *West-östliche Divan* but rather that of the travellers' geographical East. One consequence of this widening of the wanderer's world was that the sea made a conspicuous comeback in the Romantic representation of space as both a route to distant destinations and now also a feature of pristine or primordial nature on a par with mountains. With the sea, reappeared the lure of islands that had held such appeal to Defoe's and Swift's first readers.

3 Norman Backhaus, Claude Reichler, Matthias Stremlow, *Paysages des Alpes – De la représentation à l'action*, PNR 48 Rapport de synthèse of the Swiss National Science Foundation (Zürich: VDF, 2007), 59.

4 Coleridge's pantisocratic scheme of emigration to North America did not become the subject of his verse.

The literary premises of the change in outlook on this geographical nature have been traced back to the outlandish landscapes of *The Poems of Ossian* (1762) and those of sea and border ballads. The Romantics' vision of space was, however, often in the first place inspired by their own travel experiences. Except for the temporary interruptions of the French Revolution and Napoleonic Wars, travelling for pleasure and out of pure interest had become easier and more widespread since the end of the Seven Years' War. A greater number of men and women were able to travel far and wide. The lives of Romantic writers themselves testify to the fact. In 1790 the young William Wordsworth, who rightly claims to have been a "wanderer"[5] beside a confirmed rambler, thought nothing of walking most of his way through France, from Calais to the Alps, and across the Alps into Italy with a friend. Shelley and his wife, Mary, journeyed in a more leisurely and comfortable fashion by coach to Geneva and later to Italy. Their friend Byron's many travels led him to even little known countries like Albania. Like Wordsworth, all three ventured into the Alps and across some of their passes. Besides these personal experiences of new landscapes, and for more remote countries, Romantic writers also found inspiration in travellers' accounts and other writings of geographical nature that continued to be popular throughout the period. The character of the latter had changed in the course of the eighteenth century. Their focus had shifted from mapping and the inventory of natural resources to the local colour, a feature illustrated by the two most prestigious recent sources of exotic information on remote lands, Louis Antoine de Bougainville's and James Cook's reports of their overseas voyages published, the former in 1771 and the latter in 1777 and 1784. Coleridge's "The Rime of the Ancient Mariner" and Byron's South Seas tale, *The Island*, evidence this type of influence.

The Pre-Romantic period had seen topography and biogeography supplant cartography in the representation of geographical space, and Wordsworth's "Small Celandine" or the first stanzas of Shelley's "Sensitive Plant", worthy as they are of a description in a botanist's herbarium,[6] testify that poets continued to devote the same close attention to their natural surroundings as their predecessors. Under the influence

5 *The Prelude*, VI, 261, *William Wordsworth: The Major Works,* 457. Ann Radcliffe also presents her travellers as wanderers in *The Mysteries of Udolpho*, 32.

6 *Shelley: Selected Poetry*, ed. Neville Rogers, (London: Oxford University Press, 1968), 359–60, 369–71.

of naturalists like Alexander von Humboldt or Horace-Bénédict de Saussure, his contemporary and the second man to reach the summit of Mont-Blanc,[7] the study of rock formations and the measuring of the altitudes of hills and mountains contributed to giving visitors to the Alps,[8] and people in general, a more comprehensive conception of the relief of land spaces. The Romantics' interest in physical geography went still further. As will appear below, they were fascinated with its more dynamic aspects, hydrology, glaciology, climatology and meteorology. To this should be added that writers could not have been ignorant of the remarkable work of some of the same artist topographers and painters who had made the Pre-Romantics' countryside visually popular. Several of them, following in the naturalists' and first visitors' footsteps, toured the Alps: William Pars in 1770,[9] John Robert Cozens in 1776 and 1782,[10] Francis Towne and John 'Warwick' Smith in 1781, not forgetting the most famous of them all, J.M.W. Turner, whose tour of the Alps in 1802, the first of four visits, was crucial to the development of his art. It is interesting to note in the context of this study that he too had first trained and made his mark as "a topographical draughtsman and watercolourist".[11] Together these artists familiarized the public with

7 His "Voyages dans les Alpes", the first volume of which appeared in 1779, directly influenced Kant's discussion of beauty.

8 Gavin R. de Beer, *Alps and Men: Pages from Forgotten Diaries of Travellers and Tourists in Switzerland* (London: Edward Arnold, 1932). Claude Reichler and Roland Ruffieux, *Le voyage en Suisse: Anthologie des voyageurs français et européens de la Renaissance au XXe siècle* (Paris: Robert Laffont, 1998). This anthology suggests that the discovery proper of the Alps as mountains was that of the generation preceding the English Romantics.

9 William Pars travelled through the Swiss Alps in the company of Horace-Bénédict de Saussure. See Andrew Wilton, *William Pars: Journey through the Alps. Reise durch die Alpen. Voyage dans les Alpes* (Dübendorf (Zürich): De Clivo Press, 1979). He is said to have introduced the London public to the Swiss Alps (William Hauptman: *L'âge d'or de l'aquarelle anglaise*, Lausanne, Fondation de l'Hermitage, 1999, 58 and passim).

10 On his first sketching journey, John Cozens's patron was the classical scholar and antiquarian Richard Payne Knight, the author of *The Landscape, a Didactic Poem* (1794). On the second, his patron was William Beckford, the author of *Vathek*.

11 David Blayney Brown, "Turner's Grand Tour: the Alps and Switzerland in 1802", *Turner and the Alps. Exposition organisée avec la Tate Gallery, Londres* (Martigny: Fondation Pierre Gianadda, 1999), 9. See also David Hill, *Turner in the Alps: The Journey through France and Switzerland in 1802* (London: George Philip, 1992).

Fig. 5.1: William Pars, “Bridge near Mount Grimsel”.

landscapes of glaciers and rocky formations (fig. 5.1). By comparison, Oliver Goldsmith in *The Traveller* (1764) had been as little aware of mountains as Thomas Coryate, the author of *Crudities* and Shakespeare’s contemporary, and Jean-Jacques Rousseau himself seems to have been more alert to the topography of the paths that he or the characters in his novels followed than to mountain heights. The development of physical geography, concurring as it did with the spread of the taste for the sublime and the picturesque in a cultivated intelligentsia, taught people to *see* mountains and contributed to lending the third dimension of geographical space an importance it had not had previously. Naturalists and topographers could be said to have given writers, or at least suggested to them, the vocabulary they needed to depict wilder sceneries.

When it came to placing the multiple seas, the spaces of nature and the centres of social life that constituted the world of the Romantics, not to mention the itineraries leading to them, a revival of a version of the cartographic representation of space was inevitable. This cartographic basis, or should we say dimension to distinguish it from Defoe's and Swift's more literal cartographic representations, is conspicuous in the accounts of the sea and sea voyages. In the representation of the land space, it takes the form of the divisions of the world space into units mentioned above: in the distribution of these units around the world and in the references to the surfaces the units occupy, even if their contours were still ill defined. To English tourists, the Alps were essentially the western ranges from Mont-Blanc to Graubünden or Grisons and from Luzern to Lugano but not the Eastern Alps and the Dolomites. With the wanderer, the geography of superficies susceptible of cartographic representations regained the importance it had lost in the representations of the Pre-Romantic rambler's narrow universe.

The naming of identifiable places has been seen in previous chapters to be a sure signal of the presence of a cartographic dimension in the representation of geographical space and, once again, one cannot but be struck by the care with which coordinates at sea are specified and places on land are named and located in Romantic works. Wordsworth's, Byron's, and Shelley's poems teem with such references and so do Mary Shelley's *Frankenstein* and Walter Scott's novels. The degree of precision still varied depending on whether the area referred to was well known or too remote to be familiar. In the latter case, the writer was allowed some scope for imagination, even though he no longer enjoyed the freedom that Swift had taken advantage of in his satirical *Travels*. Crusoe, Gulliver, the narrator in *Joseph Andrews* or the letter writers in *Humphry Clinker* had informed the reader about their points of departure, their routes and their destinations and so do the Romantic writers for their own journeys or voyages or those of the heroes in their narratives. The geographical names are markers, as in travel accounts. They serve to define the space in which the writer traces the routes that he or his characters follow. For people sharing the same horizon of geographical discourse, a name or two are sufficient to structure an area.[12] Contemporary readers seem to have once more expected the writer to satisfy their expectations in this respect. Whether they are

12 "Nommer les lieux, c'est donner une structure à l'espace". Bernard Debarbieux, "Représentations de la haute montagne et nomination du territoire", *Représenter l'espace: l'imaginaire spatial à l'école*, eds. André, Yves, Antoine Bally et al. (Paris: Anthropos, 1989), 98.

Frankenstein's movements between Geneva, Ingolstadt, Chamonix, Scotland and Archangel or the trajectory of Coleridge's Ancient Mariner from England down to the Equator, the area of the roaring forties and, finally, up to the tropics with their typical calms in the Indian Ocean or the travels of Byron's Childe Harold from the white cliffs of southern England to Lisbon, Spain, Ionian Islands and the mountains of Albania, the travels are graphically portrayed, as they are in travel accounts. The precision is the same as that found in the descriptions of rambles. Wordsworth begins his account of a walk round Grasmere Lake by giving the name of the lake to situate it in the English Lake District. He then explains that his walk takes him along its "eastern shore" and that he follows a "rude and natural causeway, interposed / Between the water and a winding slope / Of copse and thicket".[13]

It is true that routes and movements between places are not always presented with the exactness found in these examples. Geographical indications occasionally serve simply to situate a particular space of nature with respect to better-known locations or to familiar itineraries. It must be remembered that, for areas outside Western Europe, inland topography was still much less accurately mapped than the contours of coastal areas. The Scottish Highlands remained little frequented, had few roads, and there were unexplored valleys in the Alps. In such cases, the identifiable and identified geographical location corresponded to the place where the wanderer in his quest of primordial nature turned his back on society or where human activities ceased to exert a predominant influence. On the night he climbed Snowdon with a friend—the massif was too well known to need locating— Wordsworth finds it useful to state that they started from a Welsh village called Bethkelet.[14] In "Epipsychidion", Shelley dreams of sailing from northern Italy to one of the Sporades, an island "under Ionian skies" (line 422).[15] Byron's pirates in *The Corsair* are said to sail from their own unspoilt and unnamed island to attack a Turkish force at neighbouring Coron in the south of the Peloponnese, and the runaways in *The Island* leave one Pitcairn Island, where civilisation has imprinted

13 "Poems on the Naming of Places", IV, 2–4, *William Wordsworth: the Major Works*, 203.

14 *The Prelude*, XIII, 3, *William Wordsworth: the Major Works*, 578.

15 *Shelley: Selected Poetry*, 252. See also Shelley's "Advertisement," 242.

its mark, for another smaller and less hospitable one in the Polynesian Archipelago. In the absence of an identifiable location, a special emphasis is laid on the description of a plausible topography. The quotation from Walter Scott below is a good illustration. Whether the geographical information is presented in detail or merely outlined, romantic narrative prose works and poems exhibit the same close correspondence between the linearity of the story line and the representation of journeys or voyages as trajectories on maps observed in travel fiction and travel accounts.

Rather surprisingly, as superficies, the spaces of nature often present a different character depending on whether the wanderer travels through them or comes to explore them. In the first case, the descriptions of the routes and journeys are darker and reminiscent of the experiences of older travellers. The poets' "watery plain" remains as untameable and dangerous as it was for Gulliver, Crusoe or Captain Singleton. Byron's *Childe Harold's Pilgrimage* gives a poetic definition of its treacherous nature that the satirist and novelist would have approved of:

> Roll on, thou deep and dark blue Ocean—roll!
> Ten thousand fleets sweep over thee in vain;
> Man marks the earth with ruin—his control
> Stops with the shore; upon the watery plain
> The wrecks are all thy deed, nor doth remain
> A shadow of man's ravage, save his own,
> When, for a moment, like a drop of rain,
> He sinks into thy depths with bubbling groan,
> Without a grave, unknell'd, uncoffin'd, and unknown. (IV, CLXXIX)[16]

The obstacles for mariners and ships are in fact what they had always and have always been: the hazards of the weather in Coleridge's "Rime of the Ancient Mariner" and Shelley's "Lines Written among the Euganean Hills", pirates in Byron's "Turkish Tales" and more generally rivalry between sea powers. The ships' destinations are sometimes reached and sometimes not. In *Don Juan*, the hero's ship sinks in the Gulf of Lyon in the south of France, after which a tempest nearly as fantastical as those in *Gulliver's Travels* pushes the survivors' longboat round Italy and Sicily to the Cyclades. The sea as the site of the voyagers' routes in Romantic works is a space of much suffering and death. In "The Rime

16 *Byron: Poetical Works*, ed. Frederick Page (London: Oxford University Press, 1970), 251.

of the Ancient Mariner" and Canto II of *Don Juan*, Coleridge and Byron give appalling pictures of the fates endured by the hero and his shipmates. They include scenes of thirst, hunger and death, and even, in Byron's *Don Juan*, an episode of cannibalism that would have delighted Swift.[17]

To Crusoe and the anonymous narrator of Defoe's *A New Voyage round the World*, the Pyrenees and the Andes had been obstacles that traffic and trade had to master by developing practicable crossings. Daunting as they were to the traveller, the dangers of land routes could not compare with those of the sea, however. At the turn of the nineteenth century, the main roads and paths across mountain ranges like the Alps or the Pyrenees had become busier. Owing to the reports available in autobiographical travel accounts and now guidebooks, new categories of travellers ventured over a few of their passes. But the reality was sometimes different from what maps led one to expect. Certain stretches, particularly in the maze of mountain valleys, could still have unpleasant surprises in store for the traveller. In places, the roads were little better than awkward paths and it was easy to lose one's way. While crossing the Alps on their way to Italy, Wordsworth and the friend accompanying him had this experience going down the Simplon Pass into Italy in 1790.[18] They did not realize that, for a short distance, their way lay directly "in the stony channel of the Stream". Instead, they followed a well-marked track that went in the wrong direction. Fortunately, a native they happened to meet was able to set them on their way again.[19] In fiction, Scott's Edward Waverley would hardly dream of venturing into a desolate part of the Highlands if he did not have a reliable guide:

> It was towards evening as they entered one of the tremendous passes which afford communication between the high and low country; the path, which was extremely steep and rugged, winded up a chasm between two tremendous rocks, following the passage which a foaming stream, that brawled far below, appeared to have worn for

17 See Canto II, LXXV–LXXIX, *Byron: Poetical Works*, 669–70. Théodore Géricault's famous "The Raft of the Medusa" on the same subject was painted between 1818 and 1819 about the time Byron composed *Don Juan* (1819–1824): a coincidence?

18 The Simplon Pass was opened to the traffic of vehicles fifteen years later, 9. October 1805, eleven years before Byron took this road to Italy.

19 The Prelude, Book I, 519, *William Wordsworth: the Major Works*, 463.

> itself in the course of ages … The descent from the path to the stream was a mere precipice, with here and there a projecting fragment of granite, or a scathed tree, which had warped its twisted roots into the fissures of the rock. On the right hand, the mountain rose above the path with almost equal inaccessibility; but the hill on the opposite side displayed a shroud of copsewood, with which some pines were intermingled.[20]

And even where there were decent roads in the mountains, these made for slow progress and were not without dangers as, in Ann Radcliffe's novel, M. St. Aubert and his daughter realize while travelling along the foothills of he Pyrenees:

> … St Aubert saw [the muleteer] doing homage to a cross, that stood on a rock impending over their way. Having concluded his devotion, he smacked his whip in the air, and, in spite of the rough road, and the pain of his poor mules, which he had been lately lamenting, rattled, in a full gallop, along the edge of a precipice, which it made the eye dizzy to look down.[21]

However, in poetry and fiction, the dangers of the sea routes or of the roads never discourage the wanderer in his quest of nature's thrilling sights. He knows that mountain regions have their paths and secret tracks leading to them.

Defoe's and Swift's voyagers were explorers of new worlds. The Pre-Romantic rambler was never tired of returning to beloved places. The Romantic wanderer has something of both. Fond of revisiting familiar places when he is a rambler, he becomes an explorer of a sort when he ventures in territories opened up by naturalists or previous travellers. There is however this difference that the Mediterranean places and mountain areas he visits and the exotic Asian ones he imagines are also in a way visits of sites already familiar to him. They are not true discoveries in the sense that he has heard or read about them or that they belong to his stock of classical or contemporary knowledge. Shelley knows everything there is to know about Mont-Blanc or the Bay of Naples before he sets out to go and see them. He expects to be thrilled or awed by them, without these emotions being the less genuine for all that. The analysis of the cartographic dimension of the wanderer's space

20 Walter Scott, *Waverley*, Edinburgh Edition of the Waverley Novels, vol. I, ed. P. D. Garside, (Edinburgh: Edinburgh University Press, [1814] 2007), 81.

21 *The Mysteries of Udolpho*, 30.

makes it clear that, when they are no longer spaces that, as a traveller or voyager, he has to face and cross, mountains, seas and oceans lose their forbidding character and can become sites of varied pleasures and emotional or aesthetic experiences, as rural England had been for the rambler. In the account of the walk up the Snowdon referred to above, Wordsworth describes the moment when, after a long and hard climb up by night, he and a companion emerge out of the fog to find a landscape bathed in moonlight:

> A hundred hills their dusky backs upheaved
> All over this still Ocean [of fog]
> …
> Meanwhile, the Moon looked down upon this shew
> In single glory, and we stood, the mist
> Touching our very feet; and from the shore
> At distance not a third part of a mile
> Was a blue chasm; a fracture in the vapour,
> A deep and gloomy breathing-place, through which
> Mounted the roar of waters, torrents, streams
> Innumerable, roaring with one voice.[22]

In a passage of *Childe Harold*, which echoes his predecessor's "Tintern Abbey", Byron expresses in similar terms his prepossession of the sea—a prepossession he shared with Shelley.

> And I have loved thee, Ocean! and my joy
> Of youthful sports was on thy breast to be
> Borne, like thy bubbles, onward: from a boy
> I wanton'd with thy breakers—they to me
> Were a delight; and if the freshening sea
> Made them a terror—'twas a pleasing fear,
> For I was as it were a child of thee,
> And trusted to thy billows far and near,
> And laid my hand upon thy mane—as I do here. (IV, CLXXXIV)[23]

And in Shelley's "Epipsychidion", the voyage to the Ionian Sea is pictured as an enchanting summer cruise leading the lovers to a private haven of bliss. The poet, who again dwells on the crystalline surface of

22 *The Prelude*, XIII, 45–46, 52–59, *William Wordsworth: The Major Works*, 579, 580.

23 *Byron: Poetical Works*, 251.

the Mediterranean Sea by fine weather, sees its water, not as an obstacle, as Swift or Defoe do, but as a barrier protecting his privacy and a site of soft sensuality:

> The blue Aegean girds this chosen home,
> With ever-changing sound and light and foam,
> Kissing the sifted sands, and caverns hoar … (lines 430–34)[24]

The wanderer's enjoyment of his surroundings as he moves around the spaces of nature is a feeling that Smollett's Matthew Bramble, his nephew and niece were still mostly strangers to.[25] The wanderer lives his experiences, whether delightful or painful with his whole physical being and all his senses. Romantic writers have taught us to look at, listen to, feel and smell nature as we move through specific spaces.

What precedes has made it increasingly clear that, as superficies, the Romantic wanderer's land and sea spaces of nature are extremely varied and their variety owes a great deal to their climate: one of their biogeographical aspects. They range from tropical regions to polar ones. They include seaside areas, cultivated plains shaped by man's activities, hills and mountain massifs with few or no signs of man's presence, fertile or inhospitable valleys: their geophysical aspects. The Romantic wanderer is also sensitive to the changes of seasons and of the weather and their effects on the landscape. Last but not least, there is the factor of personal taste. William Wordsworth, whose sensibility and attention to his surroundings remain essentially those of the Pre-Romantic rambler, manifests a marked preference for a nature that sustains life: for landscapes showing traces of man's presence and for a nature of fields, meadows, orchards and woods, the nature of which John Keats has given an unforgettable picture in his ode, "To Autumn". In his account of his crossing of the Alps, the author of *The Prelude* shows himself more appreciative of the binding of yellow sheaves in the sun and of a sight of a scattering of haycocks in the "wondrous Vale / Of Chamouny" than of Mont-Blanc and its glaciers.[26] His wildest natural spaces are

24 *Shelley: Selected Poetry*, 252.

25 The same can be said of Samuel Johnson, as his nearly contemporary account of his own *Journey to the Western Islands of Scotland* (1775) shows.

26 *The Prelude*, VI, 456–465, 454, *William Wordsworth: the Major Works*, 462, 461.

those he finds in the mountains and valleys of the Lake District or Wales, on whose slopes sheep graze. Unlike Shelley a few years later, he does not see the highest summit of the Alps as an imposing sight arousing fascination and terror but merely as "a soulless image on the eye". Walter Scott's depiction of Scotland in *Waverley* shows that he shares Wordsworth's preferences. On the other hand, Byron and Shelley's wife, Mary, exhibit the same fascination as the author of "Mont Blanc" with extremes in nature: the sunny nature of their happy southern landscapes and South Sea islands contrasts with the forbidding wilderness of their tempestuous seas, inhospitable polar regions and mountain massifs: a nature to all evidence unfriendly to man. On the one hand, there are the smiling Mediterranean settings of Shelley's "Epipsychidion" and the second part of his "Ode to the West Wind", not to mention certain cantos in Byron's *Child Harold* and *Don Juan*. On the other hand, there are the dark mountain scenes in Shelley's *Prometheus Unbound*, Byron's *Manfred*, and Mary Shelley's *Frankenstein*. Recent in the Gothic novel, the depiction of daunting mountains and mountain valleys was something new in poetry. This is the Mont-Blanc that Shelley opposes to Wordsworth's:

> Far, far above, piercing the infinite sky,
> Mont Blanc appears, — still, snowy, and serene —
> Its subject mountains their unearthly forms
> Pile around it, ice and rock; broad vales between
> Of frozen floods, unfathomable deeps,
> Blue as the overhanging heaven, that spread
> And wind among the accumulated steeps;
> A desert peopled by the storms alone,
> Save when the eagle brings some hunter's bone,
> And the wolf tracks her there — how hideously
> Its shapes are heaped around! rude, bare, and high,
> Ghastly, and scarred, and riven. (Lines 60–71)[27]

Infinitely contrasted are then the spaces of nature that the wanderer loves to explore, travels through or roams around.

"Mont Blanc" is a curious poem. In its discussion of the relation of the mind to the world, it comes close to denying the substantial existence of what the speaker has before his eyes while producing one

27 *Shelley: Selected Poetry*, 332.

of his most concrete geographical descriptions of a landscape.[28] One reader has defined the poem as a colloquy that the speaker, standing on a bridge over the ravine of the Arve in the valley of Chamonix, is holding with the ravine and the famous summit.[29] Because poetry is a temporal medium, and because its images are presented in the familiar sequential order, the dynamic quality of the experience could just as well be said to portray a wanderer walking up a path along the river, looking sometimes down into the ravine and sometimes at its steep slopes in the first section until he reaches a spot where Mont-Blanc suddenly appears in all its grandeur as described above. The poem contains two more geographical descriptions. After that of the summit quoted above, the speaker's meditation soars, alluding in its geophysical flight to the history of the earth (lines 71–74), to the cycle of the water from the snowfalls on the glaciers to the ocean (lines 84–87, 120–26) and to the destructive movement of the same glaciers and rock falls (lines 100–14). Could the reference to "the old Earthquake-daemon" (lines 72) and to "a sea / Of fire" (lines 73–74) be a remote and imprecise echo of Horace-Bénédict de Saussure's theory about the geological formation of alpine ranges? The Genevan naturalist had recently voiced the idea that, like volcanoes, primitive mountains—Shelley uses the word "primaeval"—exhibiting vertical granite and schist strata had been pushed upwards under the pressure of "subterranean fires". It is interesting to note that, in his descriptions, de Saussure puts the same emphasis on the forces of erosion as the poet.[30] Less memorable but just as topographically correct is finally Shelley's image at the beginning of the poem of a mountain landscape with streams rushing down the slopes to form a river (lines 1–11). Despite his claim in the last three lines of the poem that "the mind of man … creates our knowledge and makes the reality of the universe",[31] geographical spatiality forms the essence of his vision of

28 I. J. Kapstein discusses this aspect of the poem in "The Meaning of Shelley's *Mont Blanc*", PMLA, 62 (1947): 1046–60.

29 Michael Ferber, *The Poetry of Shelley*, Penguin Critical Studies (Harmondsworth: Penguin 1993), 37.

30 Henri Mallet, *Description de Genève, ancienne et moderne, suivie de la Relation abrégée d'un voyage à la cime du Mont Blanc, en août 1787* (Genève: Manget et Cherbulier, [1787], 1807), 393–402. See also de Saussure's own "Discours préliminaire", *Voyages dans les Alpes*, vol. I (Neuchâtel, 1796) xi & ff.

31 I. J. Kapstein, "The Meaning of *Mont Blanc*", 1058

what he calls "The everlasting universe of things" (line 1), and it is his vivid topographical images that the reader is most likely to remember.

However varied and contrasted they are, the spaces of nature have this in common, that they seldom show signs of human presence or activity in these Romantic representations. In the countryside of Pre-Romantic poetry, farmers or field hands are sometimes described attending to their labours. In the nature of the Romantics, and not only in the wilder natural sites like the mountains, permanent dwellers are even scarcer. It is as though they had wanted to empty the spaces of nature of any human presence. Shelley for instance dreams in "Epipsychidion" of a life alone with his beloved on an Aegean island that they would have all to themselves: a dream that indirectly recalls Robinson Crusoe's appropriation of his island. Of course, in Shelley the wish does not lead to actual appropriation but it nevertheless betrays the egocentric character of the Romantic view of the spaces of nature.[32] The few individuals that the wanderer encounters in the spaces of nature are themselves temporary visitors: lonely peasants, shepherds or hunters to whom nature grants a precarious living. Other temporary visitors are humble people left out by the recent social and economic developments. One thinks here of Wordsworth's leech-gatherer, his discharged soldier, his old Cumberland beggar or his idiot boy, who find in nature a haven that society refuses them.[33] For the second generation Romantics, under the spell of the forbidding nature of mountains, dark forests and the sea, the wild spaces of nature can also be the transitory refuge of sailors, outcasts, rebels, robbers and pirates. The notion of empty lands had undergone a complete change since Defoe. For the author of *Robinson Crusoe*, the idea was to attract permanent dwellers that would make the most of the untapped resources of empty lands. For the English Romantics, nature offers uninhabited spaces in which some scrape a living while others reap ample food for the mind. But none reside permanently in its wilder spaces.

32 In *The Magus*, John Fowles gives a modern version of the motif of the dream or of the actual ownership of islands by millionaires.

33 "Resolution and Independence", "The Discharged Soldier", "The Old Cumberland Beggar", "The Idiot Boy", *William Wordsworth: the Major Works*, 260–64, 45–49, 49–54, 67–80.

Roaming in the spaces of nature or travelling through them, the Romantic wanderer enjoys a freedom of movement that he relishes. This freedom of movements is greater as his world is larger than the Pre-Romantic rambler's. Accounts of his rambles, journeys or voyages tend to follow an appropriate pattern. They begin with the wanderer turning his back on society. The departure assumes the air of an impetuous rush or simply of a run up towards a goal, and this goal is often an elevation in the landscape, a mountain at the end of a valley or some land or island symbolizing the fulfilment of an aspiration or an escape from a curse. The sea in fact even offers the Romantics a new, seldom noted pleasure: the thrill of speed. Earlier prose accounts of sea voyages insist on the long months spent at sea. Not so in Coleridge, Byron and Shelley. Coleridge's mariner at first savours the smooth passage of his ship down the Atlantic, a pleasure conveyed by the ballad rhythm of the famous account of his departure. Like the first stanzas of "The Rime of the Ancient Mariner", the beginning of *Childe Harold* conveys an impression of swift passage from England to Portugal. Stanza XIV of the first Canto starts: "On, on the vessel flies, the land is gone".[34] In *The Corsair*, the title character is particularly sensitive to the sense of freedom that sailing gives him and his mates:

> O'er the glad waters of the dark blue sea,
> Our thoughts as boundless, and our souls as free,
> Far as the breeze can bear, the billows foam,
> Survey our empire, and behold our home. (Lines 1–4)[35]

This note is struck again and again.

On land, freedom of movement is an equally valued ingredient of the wanderer's experience of space. In "Tintern Abbey", Wordsworth explains that already as a boy, he had loved roving about the hills of the Lake District: "like a roe / I bounded o'er the mountains, by the sides / Of the deep rivers, and the lonely streams, / Wherever nature led" (lines 68–71).[36] Among hills and dales, movements may be slower but the traveller finds some rich compensation in the constant changes of scenery as previous quotations suggest. Finally, at sea and on land, what

34 *Byron: Poetical Works*, 183.

35 *Byron: Poetical Works*, 278.

36 *William Wordsworth: The Major Works*, 133.

best distinguishes the wanderer from the traveller or voyager of earlier eighteenth-century fiction is the fact that, even when he is a traveller, his mind, like the rambler's, is not fixed on his destination or the next stopping place. Like Ann Radcliffe's M. St. Aubert and his daughter on their way to Languedoc, he will stop wherever there is something to admire. Frequently he has no precise destination in view. Anyway, he has time on his hands. As Byron's restless Harold sails down the Adriatic coast or journeys through Albania, he is constantly on the lookout for nature at its "fairest in her features wild" (Canto II, XXXVII), and expresses ecstatic delight at the view of "Dark Suli's rocks, and Pindus' inland peak, / Robed half in mist, bedew'd with snowy rills" (Canto II, XLII) or at the monastery of Zitza on an eminence surrounded by mountains (Canto II, XLVIII).[37] In the spaces of nature, the wanderer's movements give him a sense of lightness or even elation, as in dreams.

If, on the whole, the cartographic dimension of the Romantic wanderer's representation of space is not the aspect that holds the attention of the reader, this is largely because it is no longer the focus of attention it had been in early eighteenth-century fiction. Owing to the wide diffusion of geographical knowledge in the public, situating places was no longer a primary concern. The shared "horizon of expectation" even allowed writers to be allusive, as Defoe had often been with the bare naming of towns and villages in his novels, well aware as he was that to name a place was sometimes sufficient to call up the vision of a particular site in his readers' minds. The case was different with relief, the third dimension of topography, whose highlighting was intimately connected with the public's developing taste for mountains, wild spaces and the sublime.

This third dimension of topography became the central aspect of the Romantic geographical representation of space, the aspect deliberately intended to draw attention, and it is in its depiction that the Romantic fascination with the extremes in nature finds its clearest expression. The shift from cartographic to chorographic representation in the depiction of the Pre-Romantic rambler's space had been a first step in the taking into account and rendering of geographical reliefs. The Romantics pushed its representation further by emphasizing the vertical and oblique lines in their descriptions in the same manner as painters and watercolour

37 *Byron: Poetical Works*, 200, 201.

artists (fig. 5.2). Ann Radcliffe in *The Mysteries of Udolpho* and Byron in *Manfred* use the adjective "perpendicular" to refer to this dimension.[38] The outcome of this development is a topographical depiction of geographical space, contrasting surfaces and altitudes, distances covered and heights attained. This trait is already evident in two of Coleridge's early poems referred to in the previous chapter, "Reflections on having left a Place of Retirement" and "Lines: Composed while climbing the Left Ascent of Brockley Coomb, Somersetshire, May 1795" and in the following contemporary quotation from Ann Radcliffe's novel:

> The aspect of the country now began to change, and the travellers soon found themselves among mountains covered *from their base nearly to their summits* with forests of gloomy pine, except where a rock of granite *shot up* from the *vale*, and lost its *snowy top in the clouds*. The rivulet, which had hitherto accompanied them, now expanded into a river; and flowing *deeply* and silently along, reflected, as in a mirror, the blackness of the impending shades. Sometimes a cliff was seen *lifting its bold head* above the woods and the vapours, that floated mid-way *down* the mountains; and sometimes a face of *perpendicular* marble *rose* from the water's edge, over which the larch threw his gigantic arms, here scathed with lightning, and there floating in luxuriant foliage.[39]

The contrast with the flat two-dimensional descriptions of William Dampier and Woodes Rogers in their travel accounts, or of Defoe, Swift, Fielding or Smollett in their fiction, is striking. Where to Matthew Bramble the Scottish Highlands only appear from a distance "heaped upon the backs of one another" like the molehills on an old map,[40] in the eyes of Captain Waverley approaching them, they "now swelled into huge gigantic masses, which frowned defiance over the more level country that lay beneath them."[41] Many if not most of the thrilling sights that Romantic writers show a predilection for exhibit these vertical and oblique lines: crags, cliffs, summits, narrow defiles, plummeting gorges, promontories and dark forests of tall pines. In their descriptions, writers multiply references to heights and depths, to snow-capped mountains and deep gorges, as well as to what happens above sea level and below the surface of the water. In the wanderer's representation of space, the

38 *The Mysteries of Udolpho*, 37, 53. *Manfred*, II, ii, 4, *Byron: Poetical Works*, 396. The word recurs in other contexts and under the pens of other writers as well.

39 *The Mysteries of Udolpho,* 37. My italics

40 Tobias Smollett, *Humphry Clinker*, 234.

41 Walter Scott, *Waverley*, 34.

Fig. 5.2: John Robert Cozens, "Chamonix and Martigny—The Aiguille Verte", known today as l'Aiguille de Dru.

sublime or picturesque are mainly a topographical, a geographical affair. The third "perpendicular" dimension lends itself to the same kinetic treatment as the rambler's chorographic landscapes. Here again, the point of view moves with the wanderer, proposing prospects—the word recurs frequently in prose works of the period—and these prospects can again be said to be grafted on the cartographic delineation of the wanderer's journeys or voyages, much as William Pars's, J.M.W. Turner's and other artists' drawings and watercolours of glaciers (fig. 5.3), waterfalls and vertiginous roads (fig. 5.4) in the French and Swiss Alps illustrate accounts of their own journeys, and modern wanderers' snapshots do their remembrances of their holiday travels. The difference is not one of kind but of quality.[42]

42 The comparison is not as far-fetched as it looks. Patrons of painters often bought pictures of places that they had visited in their travels. See Davis Blayney Brown on Turner's friend, Walter Fawkes in "Turner's Grand Tour: the Alps and Switzerland in 1802", *Turner et les Alpes 1802*, 31–32.

Fig. 5.3: William Pars, "Mer de Glace above Chamonix".

Compared to the gently undulating English landscapes of the Pre-Romantic rambler, the Romantic prospects overemphasize the spatial dimensions in such a way as to dazzle the imagination. Heights are dizzy, horizons vast and depths precipitous or unfathomable. Vertical and horizontal lines are stretched to the point of distortion, even where one hardly expects them to be so. Many of the landscapes described by that great rambler, Wordsworth, are similar to those depicted by Goldsmith or Cowper, yet the Romantic touch is there. In "Lines written a few miles above Tintern Abbey" the waters of the Wye are "rolling from their mountain-springs" (line 3) as though they were those of the Arve in the Valley of Chamonix, while the poet rests below "the steep and lofty cliffs" (line 5) that enclose the valley at the spot.[43] His native Lake District is an area not just of smiling lakes but of steep slopes, of "Not giddy, yet aerial" heights with hidden vales below, of crags, "tumultuous" brooks, "waterfalls", and even "cataracts"; of "caves" and bare hillsides.[44]

43 *William Wordsworth: The Major Works*, 131.

44 "Home at Grasmere", "The Brothers", "Tintern Abbey", "Naming of Places", "Michael", "The Ruined Cottage", *William Wordsworth: The Major Works*, 174, 155, 156, 131, 202, 203, 31.

Fig. 5.4: J. M. W. Turner, "The Devil's Bridge. St Gotthard".

Shelley's evocation of "Mont Blanc" does not mention the summit of the highest mountain in Europe until line 60 in the third section of the poem. It begins with the nearly perpendicular lines of the slopes of the valley and vertical trunks of the pines that seem to forbid its approach:

> Thus thou, Ravine of Arve—dark, deep Ravine—
> Thou many-coloured, many-voicèd vale,
> Over whose pines, and crags, and caverns sail
> Fast cloud-shadows and sunbeams: awful scene,
> Where Power in likeness of the Arve comes down
> From the ice-gulfs that gird his secret throne,
> Bursting through these dark mountains like the flame
> Of lightening through the tempest;—thou dost lie,
> Thy giant brood of pines around thee clinging … (lines 12–20)[45]

45 *Shelley: Selected Poetry*, 329.

Here, the aesthetic effect takes precedence over the geographical accuracy. By comparison, a watercolour by William Pars, the topographer, looks sober (see fig. 5.1). Even John Keats, who seldom develops a geographical vision of space and vision of movement through space, gives striking examples of vertical topography in a few poems. In *Hyperion*, a fallen Saturn is depicted sitting "deep in the shady sadness of a vale / Far sunken from the healthy breath of morn, / Far from the fiery noon, and eve's one star …" (lines 1–3).[46]

Of particular interest is again one of Shelley's descriptions of a mountain landscape. *Alastor* tells the story of the eponymous hero's journey up a valley in the Caucasus, where he meets his death. The topography of the journey depicts no less than the whole extent of the vertical dimension from ocean to mountain summits. The poet's Caucasus, which is also the setting of his lyrical drama, *Prometheus Unbound*, is imaginary because, as in the drama, the poem deals with an allegorical journey to a remote and still little known area of the world. On the other hand, the description of the valley and of the course of its river once again demonstrates the poet's familiarity with the geomorphology and topography of the upper reaches of alpine valleys. This becomes patent if, instead of following the pattern of the journey upstream, one retraces the hero's steps and pays attention to the course of the river.[47] It has its source at a place where "On every side now rose / Rocks, which, in unimaginable forms, / Lift […] their black and barren pinnacles" (lines 543–45). These two lines recall the setting of the lyrical drama. At first the water "Roll[s] through the labyrinthine dell" and "Fret[s] a path through it" (lines 541, 542). Becoming a brook struggling to find its way among the rocks (line 528), it meanders through an alpine meadow among moss and polished stones. The ravine becomes deeper, the brook reaches the timber-line, flows faster, runs down stretches where the now dense forest prevents the sunbeams lighting the surface of the water. Cataracts follow, as do caverns and whirlpools. For his part, Byron has

46 "Hyperion", Book I, *Keats: Poetical Works*, ed. H.W. Garrod, (London: Oxford University Press, 1970), 221. Keats is not so much a poet of movement as of process. One of his very few wanderers is the unhappy knight or "wretched wight" of the ballad, "La Belle Dame sans Merci", (Ibid., 350–51).

47 *Alastor*, *Shelley: Selected Poetry*, 42, 41.

beautifully caught the impression of giddy heights in *Manfred*. The first picture the reader gets of the hero is of a man standing in bad weather at the "extreme edge" of a crag, at a place overlooking a torrent and tall pines so far below that the trees look very small and he can see "clouds / Rise curling fast beneath [him], white and sulphury" between him and the bottom of the valley (I, ii, 13–15, 85–86) (fig. 5.5).[48] The scene takes place, says Byron, on the northern slopes of the Jungfrau massif, an area that was attracting numerous British tourists.

Where Swift, like William Dampier, one of his models, focused on the contours of seashores, "a small Neck of Land jutting out into the Sea, and a Creek too shallow to hold a Ship of above one hundred Tuns" on a "barren and rocky" coast,[49] Romantic writers offer seascapes. In Shelley's "Lines written among the Euganean Hills" or "Epipsychidion," and in Byron's narrative poems, the central characters sail from some land or island with a promontory, cliff, hill, or mountain, to an island or some land with a hill or mountain in the background, and perhaps even a tower visible from the shore. The coastal topography systematically opposes their vertical or oblique lines to the horizontal surface of the water on which the ship sails. In the first Canto of *Childe Harold*, the hero's goal when he leaves the white cliffs of England (XII) is not to sail to Lisbon, where he stops only briefly, but to explore the Cintra massif with its "variegated maze of mount and glen" and its "horrid crags, by toppling convent crown'd" (XVIII, XIX).[50] In *The Corsair,* Conrad sails from his "watch-tower" on the top of a perpendicular cliff, symbol of his freedom. On his way to the South Peloponnese, he sails past "the high headlands of each clustering isle" (line 593) and a "jutting cape" (line 601).[51] The same contrast between height and the vastness of the sea's surface inspires Keats's famous image of "Charm'd magic casements", those of a great house or castle, "opening on the foam / Of perilous seas, in faery lands forlorn."[52]

48 *Byron: Poetical Works*, 393, 394.

49 *Gulliver's Travels*, 85.

50 *Byron: Poetical Works*, 183–84.

51 *Byron: Poetical Works*, 286.

52 "Ode to a Nightingale", 69–70, *Keats: Poetical Works*, 209.

Fig. 5.5: William Henry Bartlett, "The Summit of the Jungfrau. Scene from *Manfred*".

Coleridge, for his part, gives a topographical combined with a dynamic description of the Ancient Mariner's ship moving away from the English coast. He also takes into account the roundness of the globe and the position of the sun:

> 'The ship was cheered, the harbour cleared,
> Merrily did we drop
> Below the kirk, below the hill,
> Below the lighthouse top.
>
> The sun came up upon the left,
> Out of the sea came he!
> And he shone bright, and on the right
> Went down into the sea.
>
> Higher and higher every day,
> Till over the mast at noon—' (lines 21–30)[53]

53 *Coleridge: Poetical Works*, 187–88.

Further down the Atlantic, south of the Cape of Good Hope, the Ancient Mariner's ship sails past towering icebergs, still another vertical element.

As the above illustrations from Coleridge, Byron, Shelley and Keats show, the cliff, the mountain and the tower, the vertical topographical dimension, may be connoted positively or negatively as either prison or refuge, threat or haven, but not so the cave, another typically romantic topographical feature. Shaped like a womb, it offers protection to solitary wanderers or lovers. Associated with the cave, the sea and the mountains are kind to man. Frankenstein's monster uses a cave as his abode in the Mont-Blanc massif. In *Don Juan*, the idyll between the hero and Haidee is born in a cave and, in *The Island*, the lovers escape their pursuers by abandoning their canoe and diving below "a black rock [that] rears its bosom o'er the spray" (IV, II, 10) in order to reach a cave by an underwater passage.[54] In *Alastor*, the cave, whose sinister "crags closed round with black and jagged arms" nevertheless opens the desired way up the Caucasus river (lines 352–69).[55] In "Epipsychidion", the island in the Sporades where the poet hopes to settle is a spot "'twixt Heaven, Air, Earth, and Sea / Cradled, and hung in clear tranquillity" (lines 457–58). Its central features are "mountain-peaks" (line 463) in the middle of the island, and a tower that "overtops the woods" (487) and dominates the sandy beaches and "caverns hoar" (line 432).[56] Only in Keats's "La Belle Dame sans Merci" is the cave a sinister place. The fairy entices the knight or "wretched wight" into a grot promising delights that turn out to be illusory and deathly.

As for "the deep", the synecdoche for the sea, it must be read literally in Romantic poetry. In "The Brothers", Wordsworth's Leonard looks down from the deck of his ship into the water and "Below him, in the bosom of the deep, / Saw mountains" (lines 58–59), the mountains of his youth at home.[57] In "The Rime of the Ancient Mariner," the appearance of "The hornèd Moon, with one bright star / Within the nether tip" (lines 210–11) heralds the period of the hero's lone suffering,

54 *Byron: Poetical Works*, 361.

55 *Shelley: Selected Poetry*, 37–38.

56 *Shelley: Selected Poetry*, 252, 253.

57 *William Wordsworth: The Major Works*, 157.

which ends when he realizes how beautiful the water-snakes swimming in the water round the ship are (lines 277–91).[58] The most striking depiction of the depth of the sea comes from Shelley. His "Ode to the West Wind" depicts the effect of an autumn storm on the vegetation below the surface of the water: on "The sea-blooms and the oozy woods which wear / The sapless foliage of the ocean". The poet adds that, under the impact of the wind on the water, they "suddenly grow gray with fear, / And tremble and despoil themselves" (lines 39–40, 41–42).[59]

The vertical dimension is not limited to the description of snow-capped summits, cliffs, cataracts, gorges, caves and sea depth. One of the above quotations has already pointed out that it extends to the sky, the sun, the moon, and the cosmos. On a hill overlooking Grasmere, Wordsworth says that he once

> thought of clouds
> That sail on winds; of breezes that delight
> To play on water, or in endless chase
> Pursue each other through the liquid depths
> Of grass or corn, over and through and through,
> In billow after billow, evermore;
> Of Sunbeams, Shadows, Butterflies and Birds,
> Angels and winged creatures …
> ("Home at Grasmere", lines 25–32)[60]

In "Tintern Abbey" the poet's meditation soars beyond the boundaries of the valley of the Wye to embrace "the light of the setting suns, / And the round ocean, and the living air, / And the blue sky …" (lines 98–100).[61] One finds an echo of this embracing vision in Shelley's "Mont Blanc" (lines 84–86), where the mountain dominates the landscape, rearing its head into the sky, a head exposed to the snows that "descend" upon it from the sky (line 131).[62] In "To a Skylark", the bird rises so high into the air that it vanishes from sight and only

58 *Coleridge: Poetical Works*, 196, 198.
59 *Shelley: Selected Poetry*, 356.
60 *William Wordsworth: The Major Works*, 174.
61 *William Wordsworth: The Major Works*, 134.
62 *Shelley: Selected Poetry*, 331, 332.

its song is heard. The cloud in the lyric of that title boasts that "The sanguine Sunrise, with his meteor eyes, / And his burning plumes outspread, / Leaps on the back of [its] sailing rack" (lines 31–33).[63] Shortly before his death, Byron's Manfred, in a movement unusual for him, lifts up his eyes to the sun and invokes the "Glorious Orb! The idol / Of early nature" (III, ii, 3–4).[64] As for Don Juan and Haidee, in their happiness

> They look'd up to the sky, whose floating glow
> Spread like a rosy ocean, vast and bright;
> They gazed upon the glittering sea below,
> Whence the broad moon rose circling into sight … (Canto II, clxxxv)[65]

This extension of the perpendicular topographical dimension is hardly surprising when one remembers the role played by the study of the sky in the development of navigation and geography. And the progress in astronomy still exerted the same fascination.

By presenting nature, not simply as defined spaces, but as a producer of these spaces, a producer of space countenancing an agricultural utilization of the soil though not its misuse, the rambler had progressively transformed walks in nature into visits paid to nature, thus making its personification more than a poetic figure of speech. He retained the scientific distancing from the natural world of the naturalist. Not so the Romantic wanderer. His extension of the topographical verticality beyond its earthly limits to include sidereal space and the depths of oceans led to his well-known celebration of it as a "power". This now fully developed personification opened out the new fourth dimension so often pointed out by literary historians as one of the most characteristic traits of Romantic poetry. This fourth dimension further underlines the status of nature as living presence in the world. Its representation, which usually involves an upward or downward gaze, finds a first expression in an early poem by Coleridge. Reaching the top of a "stony Mount", the poet exclaims while admiring the view:

63 "The Cloud", *Shelley: Selected Poetry*, 369–71; 368

64 *Byron: Poetical Works*, 403.

65 *Byron: Poetical Works*, 682.

It seem'd like Omnipresence! God, methought,
Had built him there a Temple; the Whole World
Seem'd *imag'd* in its vast circumference:
No *wish* profan'd my overwhelm'd heart.
Blest hour! It was a luxury,—to be!
("Reflections on having left a place of retirement", lines 38–42)[66]

Here, the "power" is simply God the Creator's. In the mythology of "The Rime of the Ancient Mariner", a few years later, this fourth dimension is the realm of the spirits of nature that save the title character and mysteriously bring him and the rotten carcase of his ship back to the English shore. The Polar spirit, his demons and the spirits in the air set in motion the forces of nature that punish and save the mariner. For Wordsworth, nature is a diffuse "presence":

And I have felt
A presence that disturbs me with the joy
Of elevated thoughts; a sense sublime
Of something far more deeply interfused,
Whose dwelling is the light of setting suns,
And the round ocean, and the living air,
And the blue sky, and in the mind of man,
A motion and a spirit, that impels
All thinking things, all objects of all thought,
And rolls through all things.
("Tintern Abbey", lines 94–103)[67]

In other poems, Wordsworth adds that nature acts as a spiritual nurse to man: she "embrace[s]" him and fills him with "the sense / Of majesty, and beauty, and repose, / A blended holiness of earth and sky" (lines 129, 161–63).[68] Echoing "Tintern Abbey" with this difference that nature is more a power to reckon with than a comforting influence, Byron's Childe Harold exclaims at the end of the poem:

66 *Coleridge: Poetical Works*, 107. The poem was composed in 1795. The year before, in 1795, readers could have discovered a similar feeling and reference to God in *The Mysteries of Udolpho* (36), when Monsieur St. Aubert admires the sun rising on some summits of the Pyrenees.

67 *William Wordsworth: The Major Works*, 134.

68 "Ode: Intimations of Immortality" (77–84), "Home at Grasmere" (129, 161–63), *William Wordsworth: The Major Works*, 299, 177–78.

Thou glorious mirror, where the Almighty's form
Glasses itself in tempests; in all time, —
Calm or convulsed, in breeze, or gale, or storm,
Icing the pole, or in the torrid clime
Dark-heaving—boundless, endless, and sublime,
The image of eternity, the throne
Of the Invisible; even from out thy slime
The monsters of the deep are made; each zone
Obeys thee; thou goest forth, dread, fathomless, alone.
(Canto IV, CLXXXIII)[69]

Even Shelley, albeit a confessed atheist, is not impervious to the existence of this power in nature. He senses its presence behind the white summit of Mont-Blanc, the lightning and snow that fall upon it and the winds that whirl around it. "The secret Strength of things", he calls it in "Mont Blanc", "Which governs thought, and to the infinite dome / Of Heaven is as a law ..." (lines 139–141).[70] And, in "Ode to the West Wind", he implores this same "secret Strength of things" manifested in the wind to inspire his revolutionary song to change the world: "Be thou, Spirit fierce, / My spirit! Be thou me!" (lines 61–62).[71] In both poems, this force seems to be a greater and more creative power than what he elsewhere calls Necessity. It is at such moments of what can only be called nature's epiphany that the Romantic representation of land and sea is at its most distinct from the Pre-Romantic rambler's, and most remote from the representations of Defoe, Swift and geography. Nor is it any wonder that, with this visionary dimension, travels across mountains, through forests, along or on rivers and across seas should often symbolize a meeting with the divine, and that sea voyages in particular should become an image of life, as it does in "Lines written among the Euganean Hills" or *Alastor*.[72]

69 *Byron: Poetical Works*, 251.

70 *Shelley: Selected Poetry*, 332. It is the same power that, in *Queen Mab*, Shelley calls "Spirit of Nature! all sufficing Power, / Necessity! Thou mother of the world!" (VI, 197–98), ibid., 11.

71 *Shelley: Selected Poetry*, 357.

72 Unlike the present study, which describes the cosmographic dimension of Romantic poetry as an extension of the contemporary geographic conception of space as extended substance, David Spurr has recently argued in "The Study of Space in Literature: Some Paradigms" that the cosmic dimension comes closer to Plato's

At a time when industrial development was more systematically trying to harness the forces of nature for its own purposes, Romantic writers describe them as an emanation of a supra-human, magical agency completely independent of man, who should approach it with awe or, at least, respect. They represent it exerting its power on the geographical level in innumerable ways. The wanderer's spaces of nature over which this agency presides are spaces of movement, change and process, a field of forces, hidden or apparent, constructive or destructive: a space of entropy and negentropy. One often has the impression that, like in Swift, Defoe and their successors, the cartographic plane of the representation of space is the dimension of human activity, of trade, travels, wars, politics and colonisation, whereas the third topographical dimension of the Romantic vision is that of nature's activity. The sun, the moon, storms, winds, rains and snow exert their influence in the vertical dimension of topography. One could speak of a vertical impact on the horizontal plane. Walter Scott's description at the beginning of *Ann of Geierstein* of a storm causing a landslip that cuts off the characters' path soon after their leaving Lucerne on the lake of the same name is a vivid illustration of this action of nature in movement.[73] In the wanderer's space, tempests beat down on ships and the absence of winds in the air leaves them becalmed in the middle of oceans. Storms drown men, icebergs threaten to wreck ships; the sky and the weather make plants grow, buds open to the sunshine and close when the cold and rain return; birds sing, fly and soar, and clouds scudding across the sky are split asunder, rise, curl and envelop summits in shrouds of vapour; brooks rush down slopes, hollow out passages through the hardest rock and become tumultuous cataracts; waterfalls leap down headlong from crags, seracs topple, and avalanches hurtle down and crash in valleys. Even glaciers move, a phenomenon probably brought to the naturalist's attention by the eighteenth-century small ice age. The Romantics' knowledge of, and fascination with, physical geography finds its full expression in these depictions of natural phenomena.

---

description of *chora* (*The Space of English*, ed. David Spurr and Cornelia Tschichold, SPELL 17, Tübingen: Gunter Narr, 2005, 20–22.

73 *Ann of Geierstein*, Edinburgh Edition of the Waverley Novels (Edinburgh: University Press, [1829] 2000), chapters 1–2.

Fig. 5.6: J. M. W. Turner, "The Bay of Baiae, Apollo and the Sybil".

Shelley's "Ode to the West Wind" offers a fascinating depiction of the display of these vertical forces on the geographical space and of their regulating influence. The poet juxtaposes three pictures: that of the Bay of Baiae near Naples on a summer's day bracketed between two other pictures of the great Autumn storms which, in Italy, announce the approach of winter. In the central one, gentle horizontal lines prevail. The poet reminisces how his eyes rested on the brilliant surface of the crystalline waters of the blue Mediterranean. The pumice isles seen from the shore only undulate the surface of the horizon, as they do in a painting by William Turner of the very same bay, *The Bay of Baiae, Apollo and the Sybil* (fig. 5.6). The atmosphere is peaceful. One has the impression that time has stopped. Under the sun, the blue Mediterranean

> … lay,
> Lulled by the coil of his crystàlline streams,
>
> Beside a pumice isle in Baiae's bay,
> And saw in sleep old palaces and towers
> Quivering within the wave's intenser day,
>
> All overgrown with azure moss and flowers
> So sweet, the sense faints picturing them! (Lines 29–36) [74]

74 *Shelley: Selected Poetry*, 355–56.

In the other two pictures, vertical lines dominate. The ode begins with a vision of autumn leaves falling and blown away with the seeds of the decaying vegetation when the storm comes racing from the Atlantic and sweeps over the Mediterranean and the Italian peninsula. The dead leaves "are driven, like ghosts from an enchanter fleeing", seeds are "charioted" (lines 3 and 6). Rain and lightning pelt down destroying the old vegetation. They exert the same effect on the underwater vegetation in the third picture, blurring and upsetting the blue surface of the waters. Waves "surge", and

> mid the steep sky's commotion,
> Loose clouds like earth's decaying leaves are shed,
> Shook from the tangled boughs of Heaven and Ocean … (lines 15–17)

The surface of the sea suddenly looks "Like the bright hair uplifted from the head / Of some fierce Maenad" (lines 20–21). And it is this vertical "commotion" that prepares the ground to receive the seeds that will ensure the rebirth and renewal of nature in spring. With its vertical lines, this vivid image of a storm makes one think of another famous painting by Turner, *Snowstorm— Steamboat off a Harbour* at the British National Gallery (fig. 5.7). It builds a sharp contrast with the central picture. The interplay between the horizontal lines of the second picture and the vertical and oblique lines of the first and third pictures of the storm and its effects are a typical Romantic trait of the poem, as it is of several other works, "The Rime of the Ancient Mariner" and *Frankenstein* among others. The contrast between the stillness of the summer scene and the destructive violence of the storm draws attention to the fact that the vertical lines, as in Turner's art, are not just static topographical features but dynamic lines spelling movement. It underlines the importance of the cycle of the seasons as another force, this time the same regulating one celebrated by Keats's "To Autumn", a poem about process if there is any. Process is the central force or power that is brought to bear on space. It has been abstractly defined as *natura naturans*. In fact, it is depicted in such a way as to dwarf man's activities and in particular his influence on its spaces. By comparison, Defoe's and Swift's representations of the cartographic space are static, space being an altogether passive datum, awaiting human stimulation and exploitation.

Fig. 5.7: J. M. W. Turner, "Snow Storm—Steamboat off a Harbour".

Going much further than it does in Pre-Romantic poetry, the personification of nature as a body of forces or power brings about a modification in the representation of man's attitude to it and of the wanderer's relationship to its spaces. The Pre-Romantic rambler had found health and serenity in his walks around the countryside. It had afforded him an escape from "the din / Of towns and cities". [75] In his search for an ever more authentic and pristine nature, where man is totally absent, the

75 "Tintern Abbey", 26–27, *William Wordsworth: The Major Works*, 132.

wanderer is led to recognize it as an agency that, by producing its own space with its own laws, challenges even more imperatively than in the previous period man's own production of space as appropriation. The wanderer is acutely aware of nature's otherness and of his own exteriority to it. This awareness has three consequences. First, his exploration of its thrilling sights calls for a better acquaintance with this otherness, an attitude manifest in the precise description of its phenomena and of their perception by the individual. Biogeography and physical geography prevent the representation of nature from falling into what John Ruskin was later to call "the pathetic fallacy".[76] A survival of the scientific distancing from nature, it sets limits to the endowing nature with human feelings and abilities. His quest leads the wanderer to wish this nature and its spaces to remain intact and unchanging, the reason why, as seen above, he seems to deny the native populations the right to permanent residence in its wilder spaces while reserving these spaces for his own enjoyment and that of his likes. This egoistical stance, by the way, is still reflected today in nature lovers' fight against developers. Finally, his rambles being "in general the result of his *mental* appropriation of this space considered as a site where walking is practised",[77] his appropriation is purely an act of the mind. As with the rambler, it is temporary and renewable.

In Defoe's and Swift's fake travel accounts and in Fielding's and Smollett's novels about journeys, the main characters share their two-dimensional world with crowds of people. Paradoxically, Romanticism—but it already begins with Pre-Romanticism—turns what are in fact public spaces, vast and sometimes unclaimed territories, into sites of a welcome solitude and private, or at least semi-private places of personal enjoyment. In this sense, Romantic poetry can be said to participate in, and even push to extreme, the shift of focus from the sphere of public life to that of the private life of the individual that a number of recent studies describe as an important trend in creative writing from the end of the seventeenth century onwards. At the very most, the wanderer invites a soul mate to share his solitude:

76 John Ruskin, *Modern Painters*, vol. III, Part iv, § 8.

77 Norman Backhaus, Claude Reichler, Matthias Stremlow, *Paysage des Alpes – De la représentation à l'action*, 73.

> Oh! that the Desert were my dwelling-place,
> With one fair Spirit for my minister,
> That I might all forget the human race,
> And, hating no one, love but only her!

exclaims Byron at the end of *Childe Harold* (Canto IV, CLXXVII).[78] And Shelley's *invitation au voyage* addressed to his beloved in "Epipsychidion" reads: "Emily, / A ship is floating in the harbour now, / A wind is hovering o'er the mountain's brow; / There is a path on the sea's azure floor / No keel has ever ploughed that path before" (lines 408–11).[79] No less private and personal is the wanderer's quest to transcend the experience of his senses and find in nature intimations of a supra-natural power. If it is customary to emphasize these idealistic aspects of the Romantic solitary encounter with nature, the representation of the wanderer's experience of nature as geographical space and, in particular, its vertical dimension corrects this all too positive view by drawing attention to more sombre aspects. The wanderer's solitude can prove bitter or painful, besides being comforting and inspiring. In Coleridge's ballad, nature inflicts a long and severe punishment on the Ancient Mariner for his crime against it, the killing of an albatross. The suffering may also result from mistaken expectations. English writers are unanimous in considering that any identification with and loss of self in nature is impossible. This is another point on which the examination of the representation of geographical space is explicit. Keats explores the possibility of a total merging in nature, as Shelley does in the account of Alastor's journey. They both conclude that any such attempt can only be traumatic or worse, suicidal. The knight in "La Belle Dame sans Merci" finds himself condemned for the rest of his life to haunt "the cold hillside" where he met the "fairy's child". A high degree of intimacy is possible, says "Ode to a Nightingale" but, however tempting the feeling of mysterious unity may be, a normal reaction of retreat sets in before selfhood is effaced and fusion between the subject and the object takes place.

Nor do Byron's heroes, Child Harold, Don Juan and Manfred find in nature more than brief spells of contentment and comfort for their burdened consciences. Not even Frankenstein's monster can discover a

78 *Byron: Poetical Works*, 250.
79 *Shelley: Selected Poetry*, 251–2.

permanent refuge in glaciers and the ice fields of the North Pole. As for "Stanzas Written in Dejection, near Naples", it portrays Shelley unable to find consolation from the same enchanting sunny setting that delighted him in the central part of "To the West Wind" and that Turner had painted with wonderful warm colours. The serenity of the landscape on the contrary increases his despondency. One must not expect from nature what it cannot give: a mind burdened by guilt or sorrow may find temporary relief in his incursions into nature, yet not a solution to its problems. The contact with nature is only happily rewarding for those content with the contemplation of its beauties. To expect more is to expose oneself to disappointment. The study of the wanderer's relation to the spaces of nature tends to show him as unequally happy in them and even sometimes as downright ill-at-ease in his solitude as he is in society.

A direct lesson that the wanderer's transcendentalized nature teaches man is appreciation of what it offers: gratitude for the environment, for what it yields and for its splendours and mysteries. An object of contemplation, it enhances meditations and elevates thoughts. It tells man that even seemingly barren sites are a wealth to treasure. This is one of the interesting discoveries of the Romantics. Confronted with the gigantic proportions of the three-dimensional topography of nature, the Romantic wanderer's recognition of its otherness and of his exteriority to it easily convinces him that he must be content to be a humble visitor in its spaces and a "stranger" in them. It is telling that the word should turn up under the pen of the poet who felt most at ease with nature in his wanderings, namely Wordsworth.[80] For him, man is a stranger in nature because he has become estranged from it. The idea recurs in "Tintern Abbey" (lines 50–66) and "Ode: Intimations of Immortality" (lines 81–84).[81] In these poems he gives poignant expression to the feeling of a loss and severance. The return to ordinary life is felt the more deeply as a dispossession as the expectations from nature were considerable. What then remains are the memories, which inspire the poems. The poems no longer offer simple images of nature, snapshots of landscapes, flowers or trees but records of lived experiences in and of the space of nature with its wonders:

80 *The Prelude*, I, 601, *William Wordsworth: The Major Works*, 390.

81 *William Wordsworth: The Major Works*, 133, 299.

There was a time when meadow, grove, and stream,
The earth, and every common sight,
    To me did seem
  Apparelled in celestial light,
The glory and the freshness of a dream.
It is not now as it has been of yore;—
  Turn wheresoe'er I may,
    By night or day,
The things which I have seen I now can see no more.
("Ode: Intimations of Immortality", lines 1–9)[82]

Torn between his yearning for a communion with nature and his consciousness of irreparable separation—the language is religious—unhappy in society and sometimes in nature too, although he finds it difficult to admit it, the wanderer is a typical representative of a certain elite of the time, whose approach to nature is intellectualized. As the wanderer's experience of the spaces of nature as space was seen above to solicit all his senses and bodily energies, his experience of the vertical dimension of geographical space solicits his moral being. In this he can perhaps be viewed as a remote forerunner of modern travellers in quest of authenticity rather than debased commercial exoticism.

The Romantics wanted to change man's views on the world as exclusively a space where human "living takes place" and as the scene of his production of space. They saw that such a way of looking at things must be put limits to and they did so at a time when utilitarian values were already carrying the day. Yet, although their reverential attitude to the space of nature was at variance with Defoe's, Fielding's or Smollett's, this should not conceal the fact that their representations of this space were basically the product of the same perception and conception of reality. When Coleridge, Wordsworth, Byron, Shelley, Mary Shelley and Scott viewed nature as occupying specific spaces that could be identified, located, roughly delimited and visited, and when they combined cartographic and topographic descriptions of space in accounts of their own roaming or of their characters' movements, their descriptions of landscapes simply reflected the evolution that had taken place in geographical discourse. However striking it might look and however much it might capture the attention, their emphasis on the

82 *William Wordsworth: The Major Works*, 297.

vertical topographical dimension was an offshoot of the conception of reality as extended substance, *res extensa*, inherited from the Renaissance and considered by Descartes and Newton to be rational. Does not Shelley himself, dissatisfied as he is with this way of looking at the world, refer to it explicitly in *The Witch of Atlas* when he describes a flowering lawn as "a space extended" (line 275)?[83] However striking the highlighting of the vertical topographical and fourth dimensions in the representation of the Romantic space, it amounted essentially and simply to a preference for another viewpoint. This reliance on the Newtonian and Lockean conception of reality is a reminder that, however much they tried to transcend its limitations, their contribution to the new way people came to see the world was of the same order as that of geographers, natural scientists, painters and even composers—one thinks here of Robert Schuman's overture *Manfred* and of Felix Mendelssohn's overture, *The Hebrides*, sometimes referred to as *Fingal's Cave*. Their works addressed a select elite but, through them, the message spread. One could even go so far as to say that the Romantic writers' and painters' strong delineation of mountain and sea scenery, with its emphasis on "height, depth, solitude [that] strikes, terrifies, absorbs, bewilders"—Fuseli's words[84]—did much to develop people's sense of the third dimension of topography. It played a role similar to the spatialization of the cartographic models of the first half of the century.

With a more adventurous frame of mind than the rambler's, the Romantic wanderer once again enjoyed discovering places not so well known and the poetry and fiction celebrating him did a great deal to popularize these places. Today we think of the spaces of nature as in need of protection against urban and industrial expansion and pollution. Two hundred years ago, towns and cities were small and the majority of the population lived in the country. Writers did not yet defend nature against encroaching social and economic developments. On the other hand, the literary qualities of their Pre-Romantic and Romantic representations of the space of nature, their expressive accentuation of vertical lines, their use of contrasts, sense of perspective and narrativization of space[85] in their accounts of the wanderer's movements through landscapes and of

83 *Shelley: Selected Poems*, 209.

84 Johann Heinrich Fuseli, *Lectures on Painting*, 179.

85 A concept used in analyses of nineteenth-century fiction but already applicable to the Romantic representation of space.

his encounters with the drama of geophysical forces gave visibility to, and awakened an interest in, spaces that had been deemed of little interest, either because they were too familiar or because they looked unproductive and more forbidding than attractive. Their representations of the spaces of nature invited people to discover them for themselves and, by enhancing the value of landscapes and insisting that they should be kept intact and enjoyed, they even eventually lent them an unexpected economic value as the developers of tourist resorts know too well today.

The Romantic depiction of topography also helps to understand why their world often looks so much more modern and closer to us than Robinson Crusoe's or Lemuel Gulliver's schematic cartographic ones. Romantic topography makes the spaces of nature look concrete in a way schematic cartographic representations do not. It appeals more directly to our senses, in particular our sense of sight so solicited since by photography. It is a telling illustration of the (r)evolution in sensibility that, starting with the Pre-Romantics and finding its complete expression at the end of the long eighteenth century, still influences our way of seeing the world. As a result of this evolution, Defoe and Swift may appear more remote from us than they once were but it would be wrong to conclude that their vision of the world and attitude to it no longer has any relevance to our time, far from it. In fact, in so far as the management of space is concerned, it is interesting to note that we still live today in a world torn between, on the one hand, a will to organize, control and exploit every parcel of land and of "the watery plain", not to speak of the universe, and, on the other, a desire to keep and enjoy the same mountains and seas in their original purity. We are still romantics during our weekends and holidays and live the rest of the year by the values of Defoe's materialism. At the same time, the accounts of overseas voyages of the end of the seventeenth century and early eighteenth, Defoe's in particular, and the depiction of the Romantic wanderers' roaming(s) were at one in their premonitory vision of a global world, open to enterprising travellers: for some a world open to commerce and the exploitations of the resources of the earth, and for others a world open to visitors in search of changes of scenery and culture although, unlike the Romantic wanderer, these visitors have far too often been content with cheap *dépaysement*.

Conclusion

The close connection between geography and literature examined in this study lasted as long as geography played a central role in shaping people's knowledge of the earth. It lasted as long as the shaping of this knowledge was the affair of all, professionals and amateurs, and as long as its representations appealed to the imagination of the public. Literature fell under its spell before it gained universal authority and contributed its mite to the new way of looking at the world it engendered. The modern culture of the traveller roaming around the world with map and today camera was born. For a hundred years, geography and literature interacted with each other as they had never done before and this, at a crucial moment in their respective developments. From the mapping of the globe and collection of information about remote lands that drew Defoe's and Swift's attention early in the eighteenth century to the broader and more diversified field of investigation dominated by naturalists of a hundred years later, both evolved beyond recognition but in parallel ways. Their representations of space took many forms but their kinship remained striking. There is no doubt that, by allowing a direct taking into account of the world and the present, geography played a role in severing, or at least loosening, the ties of literature with the classical tradition that had been predominant since the Renaissance. This modern dimension may well be one of the explanations behind the favour that the works possessing it have enjoyed with readers ever since.

Geography and literature parted company between the late eighteenth century and early nineteenth century, when the orientation of geography changed. Trying to define its specificity among the natural sciences, geography became less concerned with the original *geo graphein*, the description of the earth. The gathering of data yielded to the more ambitious aim of explaining natural phenomena. Poetry turned its back on this new geography. Following the example of John Keats's narrative poems, "Lamia" and "The Eve of St Agnes", Tennyson, the Pre-Raphaelites and the symbolists after them went back to the world of imagination, dreams, myths and mediaeval legends to celebrate the life of the mind and the emotions. Their lyrics sang moments in time rather than space. But there was a difference. The new poetry evidences a self-consciousness about its neglect of contemporary spatial reality absent in poems of the sixteenth or seventeenth centuries.

It is, of course, in fiction that the heritage of the close connection between geography and literature has remained most evident. Credibility based on the use of actual geographical data had hardened into one of the tenets of the art of the novel. In this, its representation of space has been largely faithful to the eighteenth-century conception of descriptive geography that inspired Defoe or Fielding. Most fiction continues to refer to known locations, to depict existing topographical milieus and to construct verbal equivalents of cartographic and/or chorographic spaces in which to map out the characters' movements. From the nineteenth century onwards, the narrativization of space in fiction has received various emphases under the influence, this time not of contemporary geography, but of literary theories like realism, naturalism or psychology as the depictions of Dickens's England, R.L. Stevenson's and Somerset Maugham's South Sea Islands and James Joyce's Dublin demonstrate. Only once was the former direct link revived: when a new descriptive branch of geography appeared in the latter part of the nineteenth century: regional geography.[1] The best illustration of its influence in English literature is Thomas Hardy's depiction of South West England or Wessex. The role played by the Fens (Norfolk) in Graham Swift's *Waterland* is a recent example.

Generally speaking, the use of geographical material to create representations of space has become as diverse as there are novelists and even works. Normally used to anchor an action in a political, economic, sociological, anthropological, or any other context, geographical layouts in novels have occasionally become quite sophisticated, multiplying referents connected with different narrative levels or changes of scenes to create effects of recognition or disorientation. Hans Castorp's train journey to the Swiss Alps in Thomas Mann's *The Magic Mountain* and the journeys of two generations of characters in Ian MacEwan's *Black Dogs* are two examples of such developments. The reference to

1 An approach usually connected with the names of Alfred Hettner in Germany and Paul Vidal de la Blache in France. For an illustration of the influence of this approach on literature, see Bernard Demont, *Représentations spatiales et narration dans les contes et nouvelles de Guy de Maupassant* (Paris: Champion, 2005), 185–187.

geographical space is so basic to fiction that authors or their editors still occasionally supply maps for remote countries or the past, confirming that such references to geography are much more than background information about the characters' scenes of action. Amitav Ghosh, the Indian writer, provides an historical map for *Sea of Poppies*, his novel about the opium trade between late eighteenth- and early nineteenth-century. Even twentieth-century meta-fiction can occasionally be said to define its status by an unexpected reliance on or, like poetry, by a blatant rejection of actual geography as Kurt Vonnegut's *Slaughterhouse Five* and John Barth's *Lost in the Funhouse* respectively show. The importance of a geographical dimension in literary representations of space is of course hardly surprising when one considers that the globe is more than ever today our "time and place", as Anita Desai says,[2] and how pressing the issue of available space is becoming and will be in the future.

Geography did not only influence the course of literature. Directly and indirectly, it had an impact on the way people looked at their surroundings and the world. As we have seen, the eighteenth-century representations of geographical space, but not only the literary ones, highlight two of the most basic concerns of the century: people's movements and their occupation of the soil. Geography was an incentive to travel and explore and little needs to be added to what has been said about the importance and diversity of these movements over the surface of the earth from overseas voyages to rambles, or about the multiplication of the means of transport—and this development was only a beginning: the steam engine was not far off. Mobility had become the affair of all classes and much eighteenth-century creative writing is a hymn to their mobility and, beyond it, to motion with the celebration of the kinetic perception of movement and natural forces. It is true that accurate assessments of distances and speeds were not the strong point of such works. More precise depictions of the time factor only came in the next century.

For their part, the eighteenth-century descriptions of man's occupation of the soil and of their actual or imagined productions and command of space led individuals to reconsider their place in society. In so far as people ceased to see themselves as simple links in a social hierarchy,

2 Anita Desai, *In Custody* (Harmondsworth: Penguin, 1984), 110.

they became inclined, partly under its influence, but not only, to think that they shared the surface of the globe with their contemporaries. Such conviction had far-reaching and conflicting consequences for mentalities. To put it simplistically, it could on the one hand favour the idea that if an individual breathed the same air as other individuals forming their society, he or she was entitled to a portion of the soil that fed them. Such a spatial view of one's place in the world sent an egalitarian message. On the other hand, the premium set on the ownership of the soil by the same view could only whet the appetite for its possession and be a source of social, economic and political tensions. It did not only promote individualism but also encourage egoism. The recurring references to discoveries and explorations of new lands, settlements, acquisitions of properties, their exploitation or neglect, to privatizations and evictions, to the increasingly strict demarcation between public and private spaces, all these confirm the scope and heat of the debate over the occupation of the soil during the century and beyond it.[3] Even the Pre-Romantics' and Romantics' disinterested gaze on nature and their claim that certain natural spaces should escape privatization are directly related to it. In this respect, the confrontation between Defoe's and Swift's views on the subject of man's production of space was an early and for its time eminently lucid statement of some of the points at issue. Not that the problem was new. It was the outcome of an already long process, whose premises went back to the important changes that had occurred in the pattern of individual lives since the sixteenth century due to political, economic, religious and educational factors. An increasing number of people were living cut off from the village roots of their ancestors and had become estranged from the cultivation of the soil. The demand for greater individual autonomy advocated in Defoe's fiction, his speaking in praise of self-reliance and Swift's reminder that reliance on others within a social order, however imperfect, was inevitable simply offered a vivid exposition of the terms of the debate over man's place in the world that creative writing went on exploring for the rest of the century.

3 A parallel can be drawn between this debate and its resulting tensions on the geographical level and the debate opposing on the virtual level of the Net today the rank and file of the those who want to keep the Net open to all and those who want to appropriate "walled gardens" in it or restrict access to it to control its use.

For literature, geography proved an invaluable instrument of concrete visualization, as it did in the visual arts and sciences. More important than its having conferred their modern shapes on the earth, its zones, seas, continents and countries was the reflection on the use and occupation of space it stimulated. It was indeed one domain of which David Hume could have also said that it "put [the minds of men] into fermentation",[4] inviting people to open their minds to the vastness, beauty and diversity of their surroundings. But it must not be forgotten that, when creative writers refigured the world imaginatively, their representations also unobtrusively showed slants ranging from the Eurocentric and nationalist biases of all eighteenth-century geography to individual preferences or prejudices.

4 David Hume, "Of Refinement in the Arts", *Essays Moral, Political and Literary*, ed. Eugene F. Miller (Indianapolis: Liberty Fund, 1985), 271.

Index